The Musculoskeletal Manual

Second Edition

The Musculoskeletal Manual
Second Edition

JACOB S. BIRNBAUM, M. D.

Grune & Stratton
Harcourt Brace Jovanovich, Publishers
Orlando   New York   San Diego   Boston   London
San Francisco   Tokyo   Sydney   Toronto

Copyright © 1986 by Grune & Stratton, Inc.

All rights reserved. No part of this publication may be reproduced or transmitted in any form or by any means, electronic or mechanical, including photocopy, recording, or any information storage and retrieval system, without permission in writing from the publisher.

Grune & Stratton, Inc.
Orlando, Florida 32887

Distributed in the United Kingdom by
Grune & Stratton, Ltd.
24/28 Oval Road, London NW 1

Library of Congress Catalog Number 86-080143
International Standard Book Number 0-8089-1796-X
Printed in the United States of America
86 87 88 89 10 9 8 7 6 5 4 3 2 1

Contents

List of Patient Handouts   *vii*
Preface   *ix*
Acknowledgments   *xi*
Fundamental Precepts of Musculoskeletal Management   *1*

1.	**Definitions and General Principles**	2
2.	**Diagnosis**	9
3.	**Therapeutics**	16
4.	**The Neck and Upper Back**	31
5.	**Neurologic Symptoms and Diffuse Pain of the Upper Extremity**	44
6.	**The Shoulder**	57
7.	**The Elbow**	74
8.	**The Wrist**	86
9.	**The Hand**	95
10.	**The Chest**	109
11.	**The Low Back**	113
12.	**The Coccyx**	148
13.	**Neurologic Symptoms and Diffuse Pain of the Lower Extremity**	150
14.	**The Hip and Thigh**	159
15.	**The Knee**	167
16.	**The Lower Leg**	206
17.	**The Ankle**	208
18.	**The Foot**	229
19.	**Pediatric Problems**	256
20.	**Running Injuries**	282
21.	**Muscle Problems**	307
22.	**Degenerative Arthritis and Osteoarthritis**	314
23.	**The Rheumatoid Diseases**	320

24.	The Crystalline Diseases	**325**
25.	Infection	**329**
26.	Osteoporosis	**334**
27.	The Temporomandibular Joint	**338**

References *341*
Index *346*

List of Patient Handouts

3-1	Using Heat and Cold	29
4-1	Your Neck: Self Care	39
4-2	Using a Cervical Collar	40
4-3	Your Neck: Posture	41
4-4	Relaxation	42
4-5	Neck Exercises	43
6-1	Shoulder Mobilization Exercises	72
7-1	Rehabilitation Exercises for Epicondylitis	83
7-2	Tennis Elbow	84
11-1	Your Low Back: Self Care	143
11-2	Your Low Back: Posture and Precaution	145
11-3	Low Back Exercises	147
15-1	Patellofemoral Pain	199
15-2	Knee Exercises	201
15-3	Ski Safety	204
17-1	Your Sprained Ankle	224
17-2	If You Keep Spraining Your Ankle	226
17-3	How to Tape Your Ankle	228
18-1	What to Look for in a Shoe	254
18-2	Taking Care of Calluses and Corns	255
19-1	Stretching Your Infant's Shoes	280
19-2	Your Baby's or Child's Shoes	281
20-1	Preventing Running Injuries	300
20-2	Stretching	301
20-3	Healing Your Running Injury	304
20-4	Decreasing Shock	305
20-5	Controlling Pronation	306
22-1	Degenerative Arthritis	318
26-1	Osteoporosis	337

Preface

There are many excellent texts and manuals available covering the diagnosis and management of severe trauma, such as fractures and dislocations, and equally fine books on the various rheumatologic, crystalline, neoplastic, and infectious conditions that affect the musculoskeletal system. However, primary care practitioners spend a fair portion of their time dealing with musculoskeletal complaints that do not fall into these categories—from tennis elbow to low back pain to children who walk pigeon-toed. Medical education, both undergraduate and graduate, has too often either ignored or lightly passed over these ubiquitous and sometimes disabling (although admittedly only rarely life-threatening) complaints. Furthermore, useful reference materials on common musculoskeletal complaints are hard to find. This manual is an attempt to remedy that. It is intended for family and general practitioners, sports medicine practitioners, family practice residents, medical students, physician's assistants, and nurse practitioners.

The text is divided into chapters primarily by body part, and the chapters are further subdivided by complaint. The method of evaluation of each complaint is then given, along with a discussion of the diagnosis, pathophysiology, and treatment of the conditions that can cause that complaint. Cross-references to other sections or chapters are often necessary, for obvious reasons.

Diagnostic techniques mentioned in the main body of the text are discussed more fully in Chapter 2 and therapeutic techniques in Chapter 3. If, for instance, antiinflammatory medication or steroid injection is advised for a certain condition, referral to Chapter 3 will give more information on the different antiinflammatory medications available and their relative advantages and disadvantages. It will also provide information on the contraindications and precautions that need to be taken into account when using steroid injections.

Lower extremity problems in runners are rapidly becoming more and more common and are discussed separately in Chapter 20. The evaluation and management of pediatric musculoskeletal problems is the topic of Chapter 19 and the final seven chapters discuss muscle problems, degenerative arthritis, the rheumatic and crystalline diseases, infections, osteoporosis and TMJ problems.

Acknowledgments

Sincere thanks must go to the many reviewers whose suggestions and criticisms were so instrumental in making this book what it is. Special appreciation to my colleagues, students and patients over the years, from whom I have learned what I know.

The manuscript was excellent'y typed by Sally Birnbaum, Mike Friedman, and Terry Shaver. Greg Smith did the illustratior.s better than I could have dreamed. Bill Hapiuk is responsible for the fine cover design. And most of all, thanks to my very lovely wife Janina B. Zukotynski, MD, who was my number one reviewer and whose patience through the long hours of work on this project made it possible; and to David and Deborah who weren't even here when I started. Any inadequacies that remain are, of course, entirely my responsibility.

This book has been organized so as to present a step-by-step approach to the diagnosis and management of musculoskeletal conditions. It is of course impossible to anticipate or write about every possible variation in symptoms, pathology, or management. Further, opinions as to proper treatment can vary among experts. As always, it remains the task of the health care provider to apply knowledge obtained from all sources to the proper care of the individual patient.

Although every effort has been made to ensure the accuracy and completeness of the drug information presented herein, neither the author nor the publisher can assume responsibility for errors or for information that became known after submission of the manuscript for publication. It remains the responsibility of the physician to check complete product information before prescribing or recommending any drug.

The Musculoskeletal Manual

Second Edition

The Fundamental Precepts of Musculoskeletal Management

1. Always check neurovascular status distal to an injury.
2. Remember that bony tenderness can mean infection, neoplasm, or significant occult trauma and must be evaluated by X-ray and whatever other techniques are indicated.
3. If more than one joint or area of the body is involved, consider that the patient may have a systemic disease.
4. Never forget that neurologic symptoms, and sometimes pain, can be due to vascular or CNS disease or to peripheral neuropathy as well as to radiculopathy.
5. Always keep in mind the possibility of referred pain.
6. Acute monoarthritis or oligoarthritis may mean a septic joint, which must be tapped to rule out infection; but never tap into a normal joint through infected skin.
7. Always ask about allergy before injecting or prescribing medication.
8. When the picture is not clear or does not progress as expected, X-rays should be obtained.
9. Differentiate periarticular from intraarticular disease.
10. Check on tetanus immunization status with any open injury.
11. Until a definitive diagnosis can be made and specific therapy instituted, treat most acute musculoskeletal injury with cold, compression, elevation, and immobilization. Never apply heat to an acute injury.
12. In children, always consider the possibilities of epiphyseal injury or that the child is battered.

1

Definitions and General Principles

A Musculoskeletal Glossary 2
    Structures 2
    Pathology 3
Articular versus Periarticular versus Referred Pain 5
Monoarthritis: Differential Diagnosis 6
Oligoarthritis: Differential Diagnosis 6
Polyarthritis: Differential Diagnosis 7
Radicular Pain 7
Management of Acute Musculoskeletal Injury 7
Chronic Pain 7
Other Principles 8

A MUSCULOSKELETAL GLOSSARY

Structures

Apophysis: A growth region where a tendon inserts into bone.

Bone: The rigid, living structure that forms the framework upon which the rest of the musculoskeletal system is built and upon which it operates. It is made up of a matrix composed primarily of collagen fibers into which are impregnated various calcium salts. Within this structure are *osteocytes,* the living cells of bone, as well as a system of nutritional pathways. Bone is a living tissue, constantly being remodeled in response to metabolic conditions and physical stresses; *osteo-*

blasts are cells that create new bone, whereas *osteoclasts* destroy bone. Covering all this, and firmly attached to it, is the *periosteum,* a sheath of dense connective tissue.

Bursa: A cystic structure that forms between surfaces that move over each other, probably to lubricate that movement.

Epiphyseal Plate: The plate of cartilage near the end of a long bone, where growth occurs in infants, children, and adolescents.

Joint: A connection between bones. The kind of joint usually thought of is a *diarthrodal joint,* which allows free movement in at least some directions. In this type of joint the opposing ends of bone are covered by *articular cartilage,* and surrounding the joint are ligaments called the *capsule.* Lining the inside of the joint capsule is the *synovial membrane,* which is epithelial tissue that produces the *synovial fluid* that lubricates the joint.

The other type of joint that is movable (though much less so than the diarthrodal) is the *amphiarthrodal joint,* which is composed of a disc of fibrocartilage between the bones, again covered by a ligamentous capsule.

Ligament: A structure made of dense regular connective tissue that connects bone to bone, around a joint or elsewhere.

Muscle: The contractile tissue that, under voluntary or involuntary nervous system control, moves and/or stabilizes the various body parts.

Tendon: Made of tissue similar to that of a ligament, but connects *muscle* to bone.

Pathology

Apophysitis: Excessive traction on an apophysis leads to inflammation. Most common are Osgood-Schlatter's Disease of the knee and Sever's Disease of the heel.

Arthritis: Inflammation of one or more joints. Can be of autoimmune, viral, septic, crystalline, degenerative, or traumatic etiology.

Avascular Necrosis: A condition, most common in childhood, in which a loss of blood supply, whether idiopathic or due to trauma or other factors, leads to death of a small or large piece of bone. Most common are Legg-Perthe's Disease of the hip and the various locations of Osteochondritis Dissecans.

Avulsion Fracture: A fracture in which only a small chip of bone is broken off, invariably at the insertion of either a ligament or a tendon. It can usually be treated the same as a third-degree sprain.

Bursitis: Inflammation of a bursa. There are two types: the superficial bursae (e.g., olecranon and prepatellar), except when infected or directly traumatized, usually present with painless swelling, while the deeper bursae (e.g., trochanteric and subacromial) present primarily with pain.

Compartment Syndrome: The increase of pressure in a fascial compartment of the arm or leg due to bleeding, infection, or tissue swelling. When acute it can lead to nerve compression and permanent neurologic sequelae if not treated emergently. See Chapter 21.

Connective Tissue Disease: One of various systemic diseases of unknown etiology. See Chapter 23.

Contusion (Bruise): A consequence of direct trauma to muscle or bone, in which there is some extravasation of blood around or in the muscle or under the periosteum.

Cramp: A painful, localized, transient spasm of muscle with a variety of possible etiologies. See Chapter 21.

Crystalline Disease: Gout and pseudogout; conditions in which the deposition of crystals in joints leads to pain and disability. See Chapter 24.

Degenerative Arthritis (or Degenerative Joint Disease): Degeneration of joint surfaces and other consequences that result from wear and tear or previous trauma. See Chapter 22.

Dislocation: An injury in which the normally articulating surfaces of a joint are displaced out of proper position. It is invariably associated with some degree of capsular ligament damage, and sometimes with a fracture as well.

Epiphyseal Injury: Slippage or crush of, or fracture through, the growth plate.

Fracture: A disruption in the continuity of bone.

Muscle Pull: See Strain.

Myositis Ossificans: A condition in which bone forms inside muscle tissue. See Chapter 21.

Nerve Compression Syndrome: The compression of a peripheral nerve, giving rise to pain and/or neurologic symptoms. The Carpal Tunnel Syndrome is the most common.

Osteochondritis Dissecans: A condition in which a small piece of articular cartilage and attached bone loses its blood supply, necroses and breaks off, and becomes a loose body in the joint. Most common in the knee, elbow, and ankle.

Pathologic Fracture: A fracture through diseased bone, often bone involved with neoplastic disease. This possibility should be considered whenever the trauma reported by the patient seems to be too minor to produce the observed fracture.

Periostitis: Inflammation of the periosteum is caused by excessive traction there by ligaments or more commonly, tendons or muscle origins. (Some degree of periosteal inflammation accompanies many cases of tendonitis, especially when at muscle origin rather than insertion: lateral epicondylitis of the elbow and some forms of shin splints, for example).

Radiculitis: Inflammation of a nerve root; usually leads to localized as well as radiating pain.

Radiculopathy: Nerve root impingement and/or inflammation that has progressed to the point where neurologic symptoms or findings are present in the areas innervated by the involved root.

Sprain: An injury to a ligament or group of ligaments. First degree sprains involve only excessive stretching, while second degree sprains are partial tears and third degree sprains are complete tears of the ligaments. Since for the most part ligaments are not placed under excessive stress with normal motion, the diagnosis of a sprain should generally only be made with a history of a specific traumatic incident.

Strain: A stretching injury to muscle. Degrees of injury are defined analogously to those in sprains, discussed above. Proper warm-up and stretching are essential for prevention.

Stress Fracture: A weakening of bone caused by increasing stress on the bone faster than it can adapt or beyond its limit of adaptation. At what point one stops calling it a stress injury and starts calling it a stress fracture is a matter of semantics; suffice it to say that a stress fracture (or severe stress injury) can, if not properly treated (mainly by desisting from the offending activity), progress in sudden and dramatic fashion to a complete through-and-through fracture. Patients with unusual or repeated stress fractures should undergo a workup to rule out underlying metabolic disease (BUN, creatinine, calcium, phosphorous, protein, acid phosphatase, alkaline phosphatase, and electrolytes). See also Chapter 20.

Stress Injury: A milder form and antecedent of a stress fracture. See above.

Subluxation: An incomplete dislocation.

Tendonitis: Inflammation of a tendon.

ARTICULAR VERSUS PERIARTICULAR VERSUS REFERRED PAIN

The majority of patients presenting with musculoskeletal pain will refer to a particular joint or joints as the source of affliction. The most basic diagnostic decision that has to be made is whether the disability is articular in origin, i.e., within the joint itself, or rather actually arises from one or more of the various structures (tendons, ligaments, bursae, muscles) that surround the joint. Location of tenderness often makes the distinction simple. Another point is that intrinsic joint disease of whatever etiology usually (but not always) gives pain upon movement of the joint in all directions, while periarticular conditions will lead to discomfort upon movement only in certain planes. The differential diagnosis of articular disease is discussed briefly in the sections that follow, and more completely in Chapters 22-25.

Note that if a relatively pain-free range of motion is found and no significant tenderness can be elicited, one must consider the strong possibility of pain referred from other structures, musculoskeletal or otherwise.

MONOARTHRITIS: DIFFERENTIAL DIAGNOSIS

The single acutely and atraumatically inflamed joint must be considered to be infected until synovial fluid analysis proves otherwise. In other words, the joint must be tapped.* The technique for this procedure is discussed in the chapters on the individual joints; and synovial fluid analysis is discussed in Chapter 2. *Never make the mistake of tapping into a normal joint through infected skin.*

A septic joint is an emergency that must be treated aggressively if serious disability is to be avoided. (See Chapter 25.)

Another common cause of an acute monoarthritis is crystalline disease, more often gout than pseudogout. These conditions are discussed in Chapter 24.

Acute exacerbations of degenerative arthritis, often precipitated by overuse, can result in a similar picture. Rheumatoid arthritis or other rheumatic diseases can sometimes begin in one joint in a fairly acute manner. (See Chapter 23.) And, very rarely, monoarthritis may signify a primary neoplasm of the joint.

OLIGOARTHRITIS: DIFFERENTIAL DIAGNOSIS

The leading possibilities are gonococcal arthritis and the arthritis associated with various viral illnesses (especially hepatitis, mononucleosis, and rubella). Rheumatoid arthritis may begin this way, especially if small joints are involved fairly symmetrically. Related diseases and rheumatic fever should also be considered, the latter especially if the arthritis is migratory. (See Chapter 23.) An exacerbation of generalized degenerative arthritis related to overactivity is another potential cause.

---

*Exceptions might be an acutely inflamed small joint in a patient with a past history of gouty attacks, or an acutely swollen first MTP joint in a patient found to have an elevated uric acid level and without a nearby break in the skin.

POLYARTHRITIS: DIFFERENTIAL DIAGNOSIS

See the discussion at the beginning of Chapter 23.

RADICULAR PAIN

Aside from the specific evaluation discussed in Chapters 5 and 13, two possible etiologies deserve consideration:
1. *Preeruptive* (or rarely noneruptive) *herpes zoster* can give a severe, superficial burning type of pain.
2. A *diabetic mononeuropathy* can produce a radicular pattern of pain, occasionally in a previously undiagnosed diabetic. Rarely, several nerves can be involved.

MANAGEMENT OF ACUTE MUSCULOSKELETAL INJURY

1. Obviously, serious injury to other organs must be considered and dealt with first.
2. Check to be sure that there is no neurovascular compromise distal to the injury.
3. Fractures and dislocations should be immobilized pending definitive treatment. The reader is referred to any of the excellent textbooks or manuals of orthopedic or emergency medicine.
4. Minor musculoskeletal injury (bruises, strains, and sprains) are treated acutely with the ICCE regime:
   Immobilization
   Compression
   Cold (Never apply heat to an acute injury)
   Elevation
5. With open injuries, remember to check on the tetanus immunization status of the victim (see Table 1-1).

CHRONIC PAIN

This of course is a subject in itself, but since the majority of these patients complain of musculoskeletal pain, it is briefly mentioned here. Patients should only be placed into this category when an extensive evaluation has failed to reveal a specifically treatable cause for his or her misery. These patients usually have pain that is quite real, but of course psychological factors are often prominent, if not in the production of pain, then certainly in the way the patient reacts

Table 1-1 Tetanus Prophylaxis

Clean wounds	
Immunized, booster within 5 years	None
Immunized, last booster more than 5 years previous	Booster[a]
Unimmunized	Course[b]
Tetanus-prone wounds (deep, extensive, contaminated, or more than six hours old)	
Immunized, booster within 1 year	None
Immunized, last booster between 1 and 10 years previus	Booster[a]
Imunized, last booster more than 10 years previous	Booster[a] 250–500 units TIG[c] in different arms[d]
Unimmunized	Course[b], 250–500 units TIG[c] in different arms[d]

[a] 0.5 cc tetanus toxoid.
[b] 0.5 cc tetanus toxoid, repeat in one month and then one year.
[c] Tetanus immune globulin, adult dose. Check on alergy first.
[d] Some authorities recommend a course of prophylactic antibiotics (penicillin or tetracycline) as well.

to it. Thus a detailed psychological evaluation, especially looking for a masked depression or secondary gain, is indicated. Beyond this, a variety of modalities are utilized in the management of these patients, including the transcutaneous nerve stimulator, hypnosis, biofeedback, relaxation techniques, and antidepressant medications.

OTHER PRINCIPLES

1. Remember that bony tenderness can mean infection, neoplasm, or significant occult trauma, and must be evaluated by X-ray and whatever other techniques are indicated.
2. If more than one joint or area of the body is involved, consider that the patient may have a systemic disease.
3. Never forget that neurologic symptoms and sometimes pain can be due to vascular or CNS disease or to peripheral neuropathy as well as to radiculopathy.
4. Always keep in mind the possibility of referred pain.
5. Always ask about allergy before injecting or prescribing medication.
6. When the picture is not clear or does not progress as expected, X-rays should be obtained.
7. Check on tetanus immunization status with any open injury.
8. In children, always consider the possibilities of epiphyseal injury or a battered child.

2

Diagnosis

History 9
Physical Examination 10
X-Rays 11
Arthroscopy 12
Myelography 12
CT Scan 12
Bone Scan 12
EMG and NCV 13
Synovial Fluid Analysis 13
Blood Tests 14

HISTORY

1. In patients who complain of musculoskeletal pain or swelling, ask about the *duration* and *consistency* of the symptom, as well as about *precipitating and relieving factors* (position, activity, mental stress, medications). Has there been any recent *overuse*?
2. Inquire as to whether there has been *any trauma* or *strain* to the involved part. If so, the specific mechanism of injury, if remembered, can be helpful in making a diagnosis. Also ask about whether there was a tearing or popping sensation at the moment of injury, how soon the part became swollen, and if there was any ecchymosis. Remember that injury in the past may not be recognized by the patient as the cause of the current affliction, and should be inquired about.
3. Especially when the neck or back is involved, ask if there is any *radiation of the pain, weakness, numbness,* or *paresthesia* distally. Conversely, neuro-

logic complaints of diffuse pain in the extremities should lead to questioning about symptoms in the neck or back.
4. It is important to inquire about *symptoms in other joints or body parts;* a positive response certainly does not prove the presence of a systemic illness such as rheumatoid arthritis or polymyalgia rheumatica, but it must lead to consideration of such a possibility and possibly to further testing. Does the patient have a previous history of arthritis or gout?
5. Ask about *locking* or restricted range of motion of a joint, and if it is a weight-bearing joint, about *collapsing* (giving way) as well.
6. Find out about *previous episodes* of symptoms in the involved area, and what diagnoses were made and the effectiveness of any treatment given.
7. Other inquiries that should be made regarding specific complaints are mentioned in the corresponding sections in the text.

PHYSICAL EXAMINATION

This is discussed in the sections in the text on individual complaints, but a few general points will be made here.

1. Observe the *functional status* of the part in question: whether the patient can walk or does so with a limp; or how well he can use his hand, etc. This is often the most important part of the examination, as it can indicate the severity of the problem and is often the main determinant of how much protection the injured part requires.
2. Always try to localize *tenderness.* Bony tenderness should bring to mind the possibility of infection, tumor, or significant occult trauma, and the area should be X-rayed and possibly even a bone scan performed. The exception to this rule would be tenderness at commonly inflamed sites of attachment of tendons and ligaments (epicondylitis, for instance), but even in these locations a lack of expected improvement and timely resolution should lead to further evaluation.
3. *Swelling* of the joint space itself must be differentiated from the soft tissue swelling of acute injury and from effusion in superficial bursae. Joint swelling implies intraarticular disease or trauma, very possibly of a significant nature.
4. Any *redness* or *warmth* implies an acute inflammatory process.
5. Test and record the *range of motion* of any involved joints. Pain in all directions of motion usually means an articular process; periarticular conditions usually limit motion in only certain directions.
6. Always check *distal neurologic* (strength and sensation) and *vascular* (pulses, temperature, color and capillary filling) *status.*
7. Of course, a complete examination will not be needed on every patient, but it is much better to do too much than too little.

X-RAYS

A few general guidelines for when X-rays should be obtained as part of the management of musculoskeletal conditions are presented here. It must be noted, however, that this remains a topic in which great variation of opinion and practice still exists.

1. Take an X-ray of any significant trauma.
2. Patients with typical presentations of common syndromes (e.g., epicondylitis, rotator cuff tendonitis, chondromalacia patellae) need not be X-rayed, but this decision should be reconsidered if the condition does not evolve and improve as expected.
3. Always take an X-ray of children who complain of musculoskeletal pain, as the possibility of a catastrophic condition is significantly higher than in adults.
4. Bony tenderness (except at typical sites of insertional strain) or deep non-localizable pain should lead to radiographs being taken.
5. In patients above the age of 50 or so, degenerative musculoskeletal conditions become more common, but then so does the possibility of neoplastic disease. (A good percentage of cancers first present with the bony pain of metastasis.) Thus the threshold for obtaining X-rays should lower with advancing age.
6. X-rays of the low back and hips involve a significant radiation exposure to fairly radiation-sensitive tissues such as the gonads and kidneys, and so criteria are somewhat different; see the discussion in Chapters 11 and 14.
7. X-ray the suspected source (neck or back) of distal radiculopathy.
8. X-rays may sometimes have a therapeutic effect, by allaying a patient's expressed or unexpressed fear of serious occult disease (especially cancer); their use for this purpose in such a situation is certainly reasonable.
9. Always check to be sure that the patient is not pregnant or possibly pregnant before taking an X-ray. This should heighten the threshold for ordering extremity X-rays, but if still indicated they should be done, using a pelvic shield. X-rays of the low back or hips should be deferred if at all possible.
10. If suspicion of neoplasm or infection is high and X-rays are negative, order a bone scan (discussed below).

ARTHROGRAPHY

This is a technique wherein contrast material is injected into a joint (most commonly the knee) in which an internal derangement is suspected. Various intra-articular conditions invisible on a plain X-ray (such as loose bodies or torn menisci) can be diagnosed by this method with a variable level of certainty. It is an invasive and fairly painful procedure, and probably should be limited to cases in which disability is great enough so that the patient would agree to surgery if indications for operation were found. As arthroscopy techniques improve, much less arthrography is being done, but it still has its place in the diagnosis of certain conditions.

ARTHROSCOPY

Using the fiberoptic arthroscope an orthopedic surgeon can visualize pathology inside a joint with much greater accuracy than previously possible, and also can often perform surgical repairs with much less morbidity and more rapid recovery than with open arthrotomy. This procedure has revolutionized the management of knee pathology especially in the diagnosis of cruciate ligament injuries and meniscal tears, and in the treatment of meniscal and articular cartilage problems, sinovial plicae and loose bodies. Arthroscopy is also being found useful in other joints, especially the shoulder.

MYELOGRAPHY

The development of CT Scanning (discussed next) has greatly reduced the use of this technique, which carries more morbidity and is not as accurate. Used to diagnose disc and other intraspinal pathology, contrast material (and sometimes air) is injected into the spinal canal after lumbar puncture, the patient is tilted into various positions and X-rays are taken.

CT SCANNING

The development of the computerized tomography scan has revolutionized diagnosis in most fields of medicine, and the musculoskeletal system is no exception. It is now the procedure of choice in the diagnosis of lumbar disc lesions, and is quite useful in accurately visualizing other areas as well, often diagnosing fractures, neoplasms and other problems that could not be well seen with plain radiographs or other techniques.

BONE SCANNING

In this technique a radioactive tracer (usually technetium-99 attached to one of various complex phosphate ions) is injected intravenously, and the whole skeleton or the part of interest is then imaged one to two hours later. Focal tracer accumulation is most often secondary to uptake in reactive bone formation, as in that associated with infection, fracture or neoplasm, but can also occur by various other mechanisms, such as inflammation of adjacent soft tissue with increased blood flow. This is an extremely useful method for detecting bony lesions of serious import long before changes are visible on X-ray, and should be considered whenever the possibility of such a lesion is being considered and the X-ray is normal.

EMG AND NCV

These techniques are often extremely helpful in differentiating the possible causes of muscular weakness, numbness and paresthesias.

Electromyography (EMG) consists of the insertion of thin electrode needles into the various muscles to be studied, and the observation of the motor unit potentials on an oscilloscope screen. Various abnormal potentials and patterns are seen with denervation of the muscle. The specific location of the pathology (root compression, compression of a specific peripheral nerve at a specific site, or a polyneuropathy) can then be surmised from the pattern of involvement of various muscles whose innervations are known. Intrinsic muscular disease or myesthenia gravis will give their own specific electromyographic patterns.

In measuring *nerve conduction velocity* (NCV) a stimulus is applied to a peripheral nerve at a specific site, and the time necessary for muscle action potentials to be picked up distally is measured. These are then compared to normal values for the age and sex of the patient; a delay implies peripheral nerve compression somewhere in between. This technique is especially useful to confirm a clinical impression of the location of compression before surgery is performed. (Some polyneuropathies slow the conduction velocity while others do not, but a comparison with other nerves will help differentiate these from compression neuropathies.)

SYNOVIAL FLUID ANALYSIS

This must be performed whenever the possibility of infection is considered. (Techniques for specific joints will be found in the corresponding chapters. Be careful never to tap into a normal joint through infected skin.) It is also the only way to confirm definitely the diagnosis of gout or pseudogout and in addition is useful in differentiating essentially noninflammatory joint diseases (such as degenerative arthritis or an effusion due to overuse) from inflammatory conditions (such as rheumatoid arthritis). Several things must be checked for in the fluid obtained.

Table 2-1 Synovial Fluid Cell Counts

	Normal	Noninflammatory	Inflammatory	Septic
Appearance	Clear	Cloudy	Turbid	Very Turbid or purulent
White cell count (per cc) (overlap is possible)	<500	500–5000	5,000–50,000	>50,000
White cell differential		Mostly monos		Mostly polys

Appearance: Normal joint fluid is straw-colored and clear. Blood indicates either significant intraarticular trauma or a bleed into a joint due to hematologic disease (assuming of course an atraumatic tap). Slightly cloudy fluid can be seen with degenerative arthritis or overuse; turbid fluid indicates an inflammatory condition; very turbid or purulent fluid is seen with infection.

White Cell Count and Differential: See Table 2-1.

Crystals: The presence of negatively birefringent crystals means gout (uric acid); positively birefringent, pseudogout (calcium pyrophosphate).

Gram Stains and Cultures: Gram stains and cultures should always be obtained. Gram stains can be falsely negative, and the cell counts seen with inflammatory and septic arthritis can overlap; if any doubt exists the patient should be treated as if he had a septic arthritis (Chapter 25) until cultures prove otherwise.

BLOOD TESTS

Sedimentation Rate: This is an excellent screening test for systemic disease in a patient with multiarea involvement. It will be increased in rheumatic fever, rheumatoid arthritis, the rheumatoid variants, SLE, and polymyalgia rheumatica, as well as with acute infection, and can also be used to follow the level of activity of these conditions. It is not increased in crystalline disease (except during severe acute attacks) or degenerative arthritis. The ESR increases with normal aging; normal values vary with the specific technique used and the laboratory performing the test. Note that this is very much a nonspecific marker, and a persistent increase may well have a significant etiology other than those discussed above.

Complete Blood Count: Whereas peripheral leukocytosis can certainly be seen with acute joint infection, a normal white count (like a normal ESR) by no means rules it out. A finding of anemia can be a clue to the presence of systemic illness such as SLE.

Rheumatoid Factor: This is a test for various antibodies of IgG. It may be negative in a fair number of patients with rheumatoid arthritis, especially when early or mild, and also is usually negative in the rheumatoid variants. Note that it may be positive in a fair percentage of patients with SLE, and in some other conditions (such as syphilis and SBE) as well. Also keep in mind that a positive rheumatoid factor in low titer can be present normally in older patients.

Antinuclear Antibodies: These are antibodies to various nuclear constituents (DNA, nucleoproteins, etc.) almost invariably found in patients with SLE. The test can also be positive in patients with scleroderma or Sjögren's syndrome, with various drug-induced lupus-like syndromes, and in about a quarter of patients with rheumatoid arthritis. The LE cell preparation is not as sensitive or as specific as the ANA test.

Uric Acid: The higher the uric acid level in the blood, the more likely the patient is to have one or more attacks of gout. The demonstration of a serum level above defined normal limits, however, does not unequivocally prove that a patient's symptoms are due to gout. See Chapter 24.

Serum Calcium, Phosphorus, Alkaline Phosphatase, Acid Phosphatase, and *Serum Protein Electrophoresis* are useful tests in detecting and discovering the cause of bone disease or lesions. Remember that normal alkaline phosphatase levels in children and adolescents are much higher than those in adults.

HLA-B27 Antigen: This marker is found in 90% of patients with ankylosing spondylitis, and in about 70% of those with Reiter's syndrome, psoriatic spondylitis or enteropathic arthritis. It can be present in normals as well.

3

Therapeutics

Rest and Immobilization 16
Heat and Cold 17
Exercises 18
Physical Therapy 19
Antiinflammatory Medications 20
Steroid Injection 22
Pain Medications 26
Muscle Relaxants 26
Psychoactive Medications and Relaxation Training 26

REST AND IMMOBILIZATION

Rest of an injured or inflamed part is a critical measure, necessary to allow healing to take place. This may mean a variety of different things, however.

1. Limitation of use of inflamed tissues is quite important. But the precise motions to be avoided must be explained to the patient, and often he or she must be given tips on how to perform daily tasks without further injuring the painful structure. For example, in lateral epicondylitis (tennis elbow), if the patient takes your admonishment "rest the elbow" to mean limiting activities requiring flexion and extension of that joint, he or she will not be avoiding stress to the inflamed tendon insertion; it is resisted extension of the wrist and the supination of the forearm that must be avoided.
2. An area may be injured acutely and require certain measures not directly related to the mechanism of injury (bedrest for a herniated nucleus pulposus, for example).

3. In the patient with neck or back pain in which emotional factors (even if nothing more than anxiety about the injury itself or its consequences) play a part, the relief of stress (resulting from time away from work, for example, or from allayment of financial worries) can give the best rest that the neck or back has had in a long time.

Immobilization of an injured part is necessary in severe injury, and often helpful in milder trauma as well. But the longer a part is immobilized, the more likely it is that the soft tissues will become stiff, the muscles around it lose their tone, and the immobilized joint lose proprioceptive sensation. These changes are often not easily reversible, and sometimes they are permanent. If an injury is known to require a certain duration of immobilization to heal properly, fine; but by all means never immobilize a joint that does not need to be immobilized (the PIP joint in the case of mallet fingers, for example), and keep the duration of splinting or casting to a minimum. Clear guidelines should be given patients about the rapid tapering of the use of mainly symptomatic immobilization (cervical collars often fall into this category, for instance); and when immobilization is discontinued, often a period of mobilization and/or strengthening exercises will be needed.

Prevention of further injury to a vulnerable structure is another measure that falls into this category of musculoskeletal therapeutics. For example, patients with recurrent ankle sprains should be advised to avoid irregular or banked surfaces, and can be taught how to tape their ankles for support; patients with a torn meniscus should be told not to take part in body-contact sports or in activities requiring sudden direction changes or squatting.

EXERCISES

There are three types of exercises to be considered.

1. *Aerobic exercise.* This is a general term to describe exercise intended to increase the patient's cardiovascular fitness. It may range from a gradual progressive program of daily walking to cycling, swimming, running, or other sports. In addition to its cardiac and respiratory benefits, it can have quite salutary psychological effects as well, often helping to relieve anxiety and depression and promoting a sense of general well-being. For these reasons it is to be recommended to all patients; in regard to musculoskeletal pathology, though, it is often specifically helpful in patients with chronic neck or back pain. Obviously, aerobic activities must be chosen that do not exacerbate the patient's primary complaint and are consistent with his or her age and general physical condition.
2. *Flexibility exercises.* These are designed to restore a normal range of motion to body parts in which disuse has led to stiffness, and to prevent muscle

pulls and tendon injuries that can result from stresses to a muscle group that is too tight. Stretching exercise programs for specific injuries and activities will be found throughout the book. Flexibility increases as a muscle is used; this is the reason that warming up is so important in the prevention of injury.
3. *Strengthening exercises.* These have a variety of objectives: to restore strength to muscles weakened by disuse secondary to injury or therapeutic immobilization; to help restore proper biomechanics in situations where lack of muscular tone contributes to malalignment and to stresses or injuries to other structures (as in quadriceps strengthening exercises for patellofemoral problems, e.g.); to prevent sprains to the ligaments that surround joints by strengthening muscular support to the joints; and to achieve a proper balance of strength between opposing muscle groups. Again, methods for specific exercises are scattered through the volume.

HEAT AND COLD

Ice application reduces inflammation, controls swelling and can help relieve muscle spasm. It is one of the most effective and important therapeutic measures in acute soft tissue injury; and it is also very helpful in treating chronic problems, especially useful after activity and during exacerbations. It is also very helpful in preventing flare ups after steroid injection.

Warmth increases muscle flexibility and joint mobility, and is therefore most useful in cases of chronic muscle tightness and in loosening arthritic joints. It is *contraindicated* in acute soft tissue injury, since it would increase swelling and subsequent morbidity.

In non-acute problems it is best to allow the patient to decide whether cold or heat gives them the best results. The immediate contrast between cold and heat application to a single area sometimes gives better relief than either modality used alone.

Cold and heat may provide some of their benefit through a counterirritant effect. Various OTC topical medications can provide some symptomatic relief through the same mechanism. Also see the discussion of physical therapy modalities in the following section.

Heat

Wet heat seems to have a somewhat better effect than dry heat. Local wet heat may be obtained in various ways (see Patient Handout 3-1). In any case, temperatures should never be high enough to cause significant pain or erythema of the skin. The patient should generally use local heat for no more than 20 to 30 minutes at a time, and no more than four times a day or so.

Patients with hand arthritis often find that dipping their hands in molten paraffin (temperature of approximately 120°F.) gives a lot of relief. (The paraffin adheres to the skin and provides a prolonged heating effect.) Set-ups are available for home use; the use of a thermometer to monitor temperature is mandatory.

Local application of heat can be hazardous in patients with decreased sensation or circulatory impairment, and should not be prescribed in these patients except under supervision. Heat is also contraindicated over areas of malignancy, and is inadvisable in acute injury (cold being the standard thermotherapy in the latter situation).

Taking a warm bath at the start or end of the day often is helpful in patients with diffuse musculoskeletal pain and/or stiffness.

Cold

If an ice bag is not available, have the patient place some ice cubes into a plastic bag and then wrap a thin wet towel around it. In general, cold should be applied for about 20 minutes at a session, four times a day or so. (In acute injuries such as sprains, however, ice packs can be applied as often as every hour or two for the first 24 to 72 hours.) In muscular pain such as in the neck or back, having a spouse or acquaintance massage the area with the ice pack may be more helpful than just placing it there. Immediate contrast with wet heat often feels good. Ice should not be placed directly on the skin.

An area may also be cooled by spraying it with ethyl chloride or flurimethane. This is very useful in the office before injection or aspiration, and also when used by a physical therapist prior to massage, stretching or other manipulation.

The use of cold is contraindicated over areas of decreased sensation or vascular compromise, in patients with vasculitis and cold-induced urticaria, and over open wounds.

PHYSICAL THERAPY

Physical therapists can be extremely helpful in the management of patients with musculoskeletal problems, especially those with back or neck pain and those being rehabilitated from injury or surgery. They can teach exercises appropriate for the patient's condition, instruct the patient on how to avoid chronic or recurrent injury to the involved area, and advise the patient on the use of other modalities such as heat and cold. They often provide significant psychological support to patients with chronic pain and disability, and give the patient whose acute problem is slow to resolve the feeling that something is being done.

A variety of additional modalities is also employed, with great variation depending on the particular therapist's training, experience and biases. They include:

Infrared. The use of radiated heat probably has no advantage over directly applied heat, and has some disadvantages (mainly with regard to cost and safety).

Diathermy. This is a method of providing "deep" heat by the use of induced electrical currents. (Directly applied heat tends to be superficial because of the insulating effect of the high water content of the dermis.) Whether whatever extra benefit occurs is worth the high cost is problematic.

Ultrasound. Mechanical oscillating waves of high frequency are produced and transmitted to the tissues. Its salutary effect is most likely due to the production of heat; a direct beneficial effect from the vibratory forces is conceivable but as yet unproven. Its comparative efficacy also remains unproven, but there are patients who swear by it.

Electrotherapy. The direct application of small currents to the body has been reported to help in acute injury. The *Transcutaneous Nerve Stimulator* (TENS) has been effective in a fair percentage of previously unresponsive chronic pain patients. Its high cost has been a drawback, but recently much cheaper, disposable units have become available and are being produced for use in the pain of acute injury as well.

Massage. There is no question that massage by a trained individual aids in pain relief and muscular relaxation. Though much is known about technique, there is little understanding of the mechanisms by which it works.

Stretching. This can be done by the patient (after proper instruction), manually by the therapist, or with the aid of traction set-ups. It is often quite helpful.

Manipulation. This potentially useful modality has been out of favor with the medical profession because of its intimate historical association with chiropractic, but there are signs that it is becoming more widely accepted. It seems clear that there would be certain situations where the application of proper physical stresses to a structure that has a disability of movement will make more sense than the administration of chemicals to the body that surrounds the structure.

ANTIINFLAMMATORY MEDICATIONS

Aside from rest, this is probably the most-used treatment modality in soft-tissue pathology. These drugs have a significant analgesic effect, but more importantly they get at the source of pain in that they decrease the inflammatory response (by affecting prostaglandin metabolism). They are thus sometimes superior to purely analgesic medications, and have the added advantage of being nonaddicting. They have the disadvantage of being fairly expensive (except for the first-line drug, aspirin) and of having some common side effects, most commonly gastrointestinal.

Contraindications and Precautions

1. These drugs should in general not be used in pregnant patients or when the possibility of pregnancy exists, and should also be avoided in nursing mothers.

2. There is cross-allergy between aspirin and some of the other nonsteroidal antiinflammatories (NSAIs); a patient allergic to any one of them (including aspirin) should not be given any drug in the group without careful consideration.
3. Some of these drugs should be used with caution (if at all) in patients with blood clotting problems and those taking anticoagulant medications.
4. All of these drugs can lead to gastrointestinal side effects, including frank ulceration. They should not be used in patients with active GI disease, and alternatives should be carefully considered in patients prone to such pathology.
5. There is increasing evidence that chronic, long-term use of drugs that interfere with prostaglandin metabolism may, by affecting kidney blood flow patterns, lead to renal failure. There seems to be increased risk in patients on diuretics and those with other chronic renal, cardiac or autoimmune problems.
6. This listing is by no means complete; always check the package insert before prescribing any of these drugs.

Choosing an Antiinflammatory Drug

Aspirin is by far the cheapest medicine in this group. (Note that while acetaminophen is just as good an analgesic and antipyretic, it does not have antiinflammatory action.) Aspirin does, however, have some disadvantages:

1. To main antiinflammatory blood levels, it must be taken at least four times a day (though long-acting preparations are now becoming available as well).
2. It has more gastrointestinal side effects than some (but certainly not all) of the other NSAIs. This may be ameliorated to some extent by using enteric-coated aspirin or preparations which have added antacid. See Table 3-1. Other salicylate molecules are asserted to have significantly less gastrointestinal side effects as well: Some of these are listed in Table 3-2.
3. Because it is a widely used medication available without a prescription, some patients will not believe it is going to work as well as a more exotic (and expensive) prescription. Many patients can be educated out of this bias, but some cannot. This may be unfortunate, but that is the way it is.
4. Note that the contraindications to aspirin (including allergy and pregnancy) can also be contraindictions to other nonsteroidal antiinflammatories.
5. See Table 3-3 for a comparison of some of the advantages and disadvantages of the various nonsalicylate NSAIs available. Some of these drugs are not approved for use in children under age 12. Although they do not profess to be better than aspirin, there is no question that some patients respond idiosyncratically better to some antiinflammatory medications than to others; in a patient who is not benefitting sufficiently, it is reasonable to try the available medications one at a time until an acceptable level of relief is obtained. Several new drugs in this class are being introduced and marketed as analgesics.
6. *Steroids,* when administered orally, can have a variety of side effects, espe-

Table 3-1 Types of Aspirin Preparations.[a]

Plain aspirin (many)
Enteric coated aspirin (Ecotrin Duentric Coated Aspirin (Smith Kline & French), others)
Aspirin with antacid (Ascriptin (Rover), Bufferin (Bristol-Myers), Arthritis Strength Bufferin (Bristol-Myers))
Time release aspirin
Prescription aspirin (Easprin (Parke-Davis) 975 mg, adult dose IOID)
Aspirin with other ingredients
(Many such preparations are available; patients should be advised to use them only if the therapeutic effect of the added ingredients (caffeine, antihistamines, etc.) is desired.)

[a]These preparations are available over-the-counter except as noted. This chart is for the reader's convenience, only and the listing is by no means complete. Neither recommendation of the specific products listed nor criticism of those not listed is implied. Always check complete product information before prescribing or recommending any drug.

cially when administered for prolonged periods. Aside from occasional use in a rapidly tapering manner in patients with acutely herniated discs, they are not recommended.

STEROID INJECTION

Steroid injection (see Table 3-4) is often an extremely helpful measure; it deposits potent antiinflammatory medication at the site of pathology, and systemic side effects are usually insignificant. It is not without its deleterious effects and risks, however. The points that are made in the following sections must be considered and observed religiously if problems (sometimes quite serious) are to be avoided.

Table 3-2 Examples of Nonaspirin Salicylates[a]

Product	Ingredients[b]	Recommended adult dosage
Arthropan liquid (Purdue Frederick)	Choline salicylate	i tsp q4h prn
Disalcid (Riker)	Salicylsalicylic acid	ii TID or iii BID
Trilisate (Purdue Frederick)	Choline salicylate, Mg salicylate	i to iii BID (of 500 mg size)

[a]This chart is for the reader's convenience anly, and the listing is by no means complete. Neither recommendation of the specific products listed nor criticism of those not listed is implied. Always check complete product information before prescribing or recomending any drug.
[b]The choline salicylate and salicylsalicylic acid preparations are reputed to have a lower incidence of gastrointesinal side effects.

Table 3-3 Examples of Nonsteroidal Antiinflammatories[a]

Drug	Usual adult dosage	Comment
Indomethicin (INDOCIN; Merck, Sharp and Dohme) 25 mg, 50 mg 75 mg. time-release 50 mg. rectal suppos.	Usually 25 mg TID or 50 mg TID or 75 mg qd or BID	Seems to be more potent than some of the others, and useful for acute problems High incidence of GI upset, and fairly high incidence of CNS problems, especially headache.
Sulindac (CLINORIL; Merck, Sharp and Dohme) 150 mg, 200 mg	150 mg BID or 200 mg BID	BID dose convenient, less GI upset than indomethicin (and possibly aspirin) Seems fairly potent
Naproxen (NAPROSYN; Syntex) 250 mg, 375 mg, 500 mg	250 mg BID to 500 BID	BID dose convenient Approved as analgesic by FDA Relatively good potency and low incidence side effects
Piroxicam (FELDENE; Pfizer) 10 mg., 20 mg	20 mg qd	Once a day dose very convenient
Diflunisal)DOLOBID; Merck Sharp or Dohme) 250 mg, 500 mg	250 or 500 mg q 8-rh	Marketed as analgesic
Ibuprofen (MOTRIN; Upjohn; Rufen, Boots) 400 mg, 600 mg,	400 to 600 mg TID or QID	Seems to work fairly well in more chronic problems Probably less GI effects than ASA Available OTC in 200 mg strength: ADVIL (Whitehall) NUPRIN (Bristol-Myers)
Fenoprofen (NALFON; Dista) 200 mg, 300 mg, 600 mg	300 or 600 mg QID 200 mg q4–6h for pain	Approved as analgesic by FDA
Tolmetin (TOLECTIN; McNeil) 200 mg, 400 mg	400 mg TID or QID	
Meclofenamate (MECLOMEN; Parke-Davis) 50 mg, 100 mg	50 or 100 mg TID or QID	Reported to have a high incidence of diarrhea
Nonsteroidal antiinflammatories marketed as analgesics			
Naproxen sodium (ANAPROX; Syntex) 275 mg	Up to 5 a day in divided doses	Do not use with naprosyn Marketed as analgesic
Phenylbutazone group			
Phenylbutazone (BUTAZOLIDIN, BUTAZOLIDIN ALKA; Geigy; AZOLID,	Up to 600 mg per day in divided doses	Real though small risk of aplastic anemia; though potent and cheap, use not recommended.

Table 3-3 *(continued)*

Drug	Usual adult dosage	Comment
AZOLID-A, USV; available generically) 100 mg With or without antacid		
Oxyphenbutazone (OXALID, USV; TANDEARIL, Geigy) 100 mg	Same as for phenylbutazone	Same as for phenylbutazone; not recommended
	Colchicine	
Colchicine 0.6 mg	See package insert	Classically used in acute gout Use with caution

*a*This chart is for the reader's convenience only, and the listing may not be complete. Neither recommendation of the specific products listed nor criticism of those not listed is implied. Always check complete product information before prescribing any drug. Most of htese drugs are best taken with food or antacid.

Precautions before Injection

1. Always obtain *informed consent.* The patient must be told of your judgment of the chance of a good result from injection and of alternative treatments available. But even more important, he or she must be aware of the risks of, and possible deleterious effects from, the procedure (listed below).
2. Be sure the patient is *not allergic* to anything you will be injecting. The local anesthetic will much more often be a problem than the steroid.
3. *Resuscitation* equipment and expertise must be rapidly available.
4. Check the package insert for other precautions and contraindications relative to the drug you are using.

Risks of Injection

(All of these are very rare, but you and the patient must be aware of them.)

1. Introduction of *infection.*
2. *Needle trauma* to a nerve or vascular structure. (Know where you are injecting, and always aspirate before injecting.)
3. *Allergic* reaction.

Possible Deleterious Effects of Injection

1. Steroid injections near tendons can weaken them, especially if repeated often. Use of the tendon can then lead to rupture.
2. Intraarticular instillation of steroid can accelerate cartilage deterioration and

STEROID INJECTION

Table 3-4 Examples of Injectable Steroids[a]

ARISTOSPAN	(Triamciolone hexacetonide; Lederle) 5 mg/ml or 20 mg/ml
CELESTONE SOLUSPAN	Betamethasone[b] sodium phosphate, betamethasone[b] *acetate; Schering)* 6 mg/ml
DECADRON	*(Dexamethasone*[b] sodium phosphate; Merck, Sharp, & Dohme) 4 mg/ml or 24 mg/ml
DEPO-MEDROL	(Methylprednisolone acetate; Upjohn) 20 mg/ml or 40 mg/ml or 80 mg/ml
HEXADROL	(Dexamethasone[b] sodium phosphate; Organon) 4 mg/ml or 10 mg/ml
KENALOG	(Triamcinolone acetonide; Squibb) 10 mg/ml or 40 mg/ml

[a]This chart is for the reader's convenience only, and the listing may not be complete, Neither recommendation of the specific products listed nor criticism of those not listed is implied. Always check complete product information before injecting any drug.

[b]Dexamethasone and betamethasone are considered several times more potent than triamcinolone and methylprednisolone.

thus degenerative arthritis, again especially if repeated often. Short-term benefit may to some extent be at the price of long-term detriment.

3. The injection of steroid can on rare occasions lead to systemic side effects including increased insulin requirements in brittle diabetics, and menstrual disturbances.

Technique of Injection

1. *Clean* the area. If a joint is to be entered, complete sterile technique must be used throughout.
2. A topical spray of *ethyl chloride* immediately before injection is quite helpful.
3. Unless deep injection into a large joint such as the knee is being done, *mix* the steroid with local anesthetic (without added epinephrine) in the same syringe. The local will not only make the shot hurt less, but the relief noted by the patient after injection will tell you that the medication is in the right place.
4. *Never* inject under pressure; most commonly this means that the needle is in a tendon, and significant degeneration of that structure can result. And do not even inject *around* a major weight-bearing tendon such as the Achilles or patellar.
5. *Never* inject where there is even the remotest possibility of infection. If tapping a joint, do not inject steroid if the synovial fluid is anything but perfectly clear.

Instructions to the Patient after Injection

1. The patient should place ice packs on the area for 20 minutes four times a day for two days. This helps a great deal to prevent the flare of pain

which often occurs soon after injection (possibly due to the crystallization of some of the steroid suspension).
2. If injection was near a tendon, sudden ballistic stress on that tendon (as in throwing a ball or swinging a racquet) should be avoided for several weeks.
3. The shot should not be considered a cure-all; the painful structure should still be rested, and other measures such as thermotherapy and oral medication should be used as appropriate.

PAIN MEDICATIONS

When possible, acetaminophen or aspirin or one of the other nonsteroidal antiinflammatory medications should be used. However, there will obviously be cases in which the analgesic effect of these drugs will be insufficient. In these situations, or when antiinflammatory drugs are contraindicated and the pain is too severe to be controlled by acetaminophen, one of a variety of analgesic medications can be prescribed. These more potent analgesics are more likely to result in various side effects, and also can lead to dependence if overused. But keep in mind that the relief of pain is quite important in breaking the pain → muscle spasm → pain cycle which so often prolongs soft-tissue distress, and so should be considered as a measure to help bring about resolution of the episode, as well as one to provide temporary relief from discomfort.

MUSCLE RELAXANTS

Whatever muscle-relaxation effect results from the use of these medications is probably due largely to a central nervous system depressive effect or mild analgesic effect; it is probably not worth the cost and the risk of adverse reactions. However, some patients who have taken these drugs in the past will request them, and other patients will respond better to (or at least be more receptive to the prescription of) one of the combination drugs listed in the second part of Table 3.5 than to simple analgesics.

If minor tranquilizers such as one of the benzodiazepines are considered for use as "muscle relaxants," it should be recognized that they probably also exert much of their action on the central nervous system, and the decision as to whether to prescribe them should be based on a judgment as to whether their central nervous system effects are desirable in the individual case.

PSYCHOACTIVE MEDICATIONS AND RELAXATION TRAINING

Minor tranquilizers (sometimes in combination products with analgesics) can be useful in cases where anxiety is felt to be a factor in the initiation or the persis-

Table 3-5 Examples of Muscle Relaxants[a]

Drug alone	
Drug	Usual adult dosage
FLEXERIL (cyclobenzaprine HCl, Merck, Sharp & Dohme) 10 mg	i TID (maximum 2–3 weeks)
NORFLEX (orphenadrine citrate, Riker) 100 mg	i BID
PARAFLEX (chlorzoxazone, McNeil) 250 mg	i-ii TID–QID
RELA (carisoprodol, Schering) 350 mg	i QID
ROBAXIN (methocarbamol, Robins) 500 mg, 750 mg	6 gm/day acutely, then 4 gm day (always in divided doses)
SOMA (carisoprodol, Wallace) 350 mg	i QID

Combinations with analgesics		
Drug	Ingredients	Usual adult dosage
NORGESIC (Riker)	Orphenadrine citrate 25 mg, aspirin 385 mg, caffeine 30 mg	i-ii TID–QID
NORGESIC FORTE (Riker)	Exactly double norgesic	½-i TID–QID
PARAFON FORTE (McNeil)	Chlorzoxazone 250 mg., acetamenophen 300 mg	ii QID
ROBAXISAL (Robins)	Methocarbamol 400 mg, aspirin 325 mg	ii QID
SOMA COMPOUND (Wallace)	Carisoprodol 200 mg, aspirin 325 mg	

[a] This chart is for the reader's convenience only, and the listing may not be complete. Neither recommendation of the specific products listed nor criticism of those not listed is implied. Always check complete product information before prescribing any drug. All of these drugs can cause drowsiness.

tence of the patient's symptoms, or where the patient's reaction to the disability or to the prescribed therapy results in tension that complicates the situation. Their routine use in every patient with back or neck pain is to be condemned, however, and when they are prescribed it must be with a full awareness of the potential for interaction with alcohol and other drugs, for abuse, and for addiction.

Antidepressants are often helpful in patients with chronic neck and back pain, in whom depression is so often a contributing factor. The diagnosis and treatment (by pharmacologic or other means) of anxiety and depression is of course a branch of medicine in itself and is beyond the scope of this book.

The teaching of *relaxation techniques* can be the most important single therapeutic measure in patients with chronic muscle-tension states; some are discussed in Chapter 4. *Biofeedback training* is one method of teaching relaxation that seems to work well in a large number of patients and is becoming increasingly popular.

PATIENT HANDOUT 3-1

Using Heat and Cold

Heat and cold are often very helpful measures that you can use at home to cut down on pain and spasm. If you have an acute injury, you should apply *cold* to the area for the first 48 hours or so. Otherwise, use whichever one (heat or cold) gives you the most relief. Often using *contrast* treatments (that is, heat immediately followed by cold or vice versa) is more beneficial than either one alone.

HEAT

If you ache or are stiff *all over,* you will find that a nice warm bath for 20 minutes or so first thing in the morning and before bed in the evening helps loosen your muscles and joints and relieves some of the pain.

If you hurt in a *specific area,* try applying heat to that area for twenty minutes or so four times a day. Never allow it to get so hot that it is painful or that the skin gets red. A heating pad is all right, but most people feel that *wet* heat is better. You can obtain wet heat in several ways:

By draping a towel over the sore area and letting the hot shower hit it (especially useful if it's your neck that hurts).

By soaking towels in hot water and applying them to the painful area. (The problem is that they tend to cool off quickly).

By getting a heating pad specially made for wet heat. (*Never* allow a regular heating pad to get wet).

By wrapping a wet towel around a hot water bottle.

By using a *hydrocollator,* which is a silicate gel pack that can retain heat for a prolonged period after being immersed in hot water. They are available in drug stores in a vareity of shapes designed for specific body areas.

COLD

Never apply ice directly to the skin. If you don't have an ice bag, you can put some ice cubes in a plastic bag and then wrap the whole thing in a towel.

Apply the ice pack to the painful area for no more than 20 minutes at a time. If the pain is a long-term problem, do this four times a day; if you have just *injured* an area, do it every hour or two for the first 48 hours after injury.

> *Never* use heat *or* cold over areas that are numb or that have poor circulation, or over open wounds. If you are diabetic, be sure to get your doctor's approval before starting.

4

The Neck and Upper Back

Evaluation 31
    History 31
    Examination 33
Neck and Upper Back Pain 33
    Acute Torticollis 33
    Neck Pain Syndrome 34
Neck and Upper Back Trauma 37
    Acute Strain of Specific Areas 37
    Cervical Sprain 37

Conditions including *neurologic symptoms* and/or *diffuse pain in the upper extremities,* even if originating in the neck, are covered in Chapter 5. *Wryneck in infants and children* is discussed in Chapter 19. Further details on therapeutic suggestions can be found in Chapter 3.

For the anatomy of the cervical spine, see Figure 4-1.

EVALUATION

History

Inquire about the *duration* and *location* of pain. Recent sudden onset of unilateral muscular pain and spasm without any known trauma is called ACUTE TORTICOLLIS.

Ask if there has been any *acute trauma or strain*. An acute strain of one area

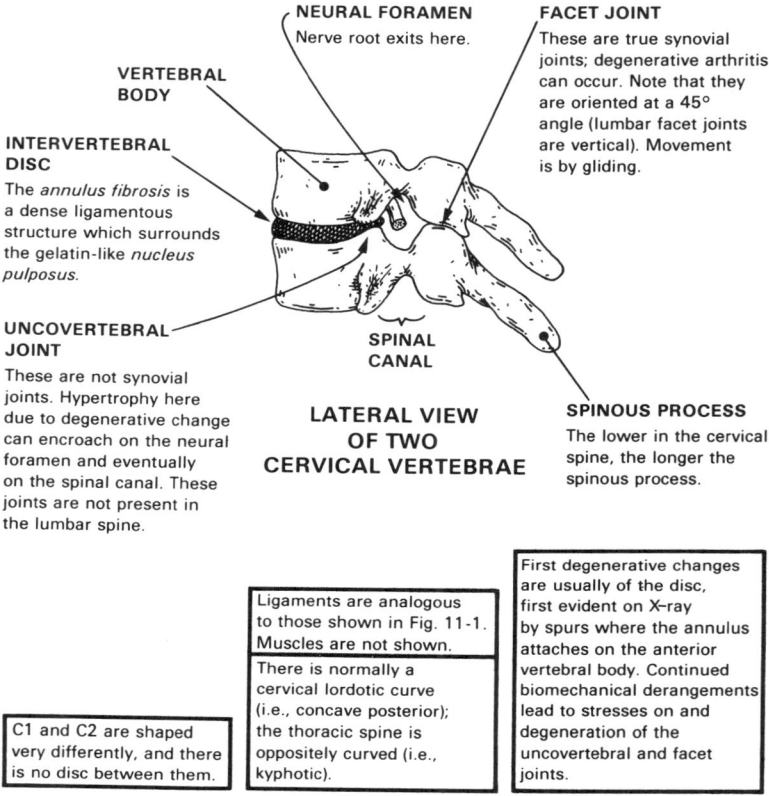

Fig. 4-1. Anatomy of the cervical spine.

is discussed on page 37, whereas a CERVICAL SPRAIN refers to severe generalized pain such as in a whiplash injury.

Is there any *radiation* of pain down the arms, or *weakness, numbness,* or *parethesias* in the arms? If so, see CERVICAL RADICULITIS in the next chapter. Note that bilateral symptoms, which may or may not also involve the lower extremities, should immediately raise suspicion of a cervical chord compression syndrome, either due to a cervical disc protrusion, a fracture, or sometimes neoplasm. (With a disc protrusion neck flexion usually exacerbates the symptoms.) Immediate cervical immobilization and neurosurgical consultation is imperative.

If none of the above apply, see the general discussion of neck and upper back pain under the heading NECK PAIN SYNDROME.

Inquire about *precipitating factors.* Especially ask about *tension, stress,* and *postural factors* such as close work with the neck bent over, sleeping on too many pillows, or lying with the neck flexed while reading or watching TV.

NECK AND UPPER BACK PAIN

Also ask about *relieving factors*.
Find out about *previous neck problems*, diagnoses, X-rays, and treatments.

Examination

Check to be sure that adenopathy or mass is not the source of the pain.

Look for *sites of tenderness*, especially over the spine; and also specifically look for *trigger points*, which are small spots of muscle spasm and are extremely tender. Chronic pain associated with fatigue, stiffness and multiple trigger points is called FIBROSITIS and is covered in Chapter 21.

Observe which *directions of motion* are restricted and/or painful: flexion (i.e., forward; normal 45°); extension (normal 55°); left and right rotation (normal 70° each way); and left and right lean (normal 40° each way). Check *voluntary* ROM; do not push.

Check *sensation, strength, reflexes,* and *pulses* in the arms and if indicated, in the legs as well.

Remember that neck pain can be due to serious intracranial or intrathoracic disease, as well as to neoplasm in the neck itself. This can be due to pain from the lesion itself, or to secondary muscle spasm.

NECK AND UPPER BACK PAIN

Acute Torticollis (Wryneck) *(Very Common)*

Diagnosis

There is a history of sudden onset (usually on awakening) of pain on one side of the neck with deviation of the neck to that side.

Often some premonitory twinges were noted before the acute event.

Frequently, localized exposure to cold (e.g., sleeping in a draft or sitting by an open car window) or prolonged positioning of the neck in an unusual position occurred just prior to the onset.

On examination, tenderness is usually diffuse on the involved side with palpable spasm, and motion away from the involved side is limited.

With repeated episodes, an X-ray should be done to rule out underlying pathology.

The condition usually resolves in a few days.

Pathophysiology

Localized muscle spasm occurs due to muscle fatigue, environmental factors, or nerve irritation.

Treatment

A *soft cervical collar* for a few days helps to "unload" the tight muscle and support the neck in a neutral position.

Intermittent *heat* to the involved area is helpful in reducing pain and spasm; sometimes ice feels better.

Analgesics should be prescribed not only for temporary relief, but to help break the pain/spasm cycle.

Neck Pain Syndrome *(Extremely Common)*

Diagnosis

The pain may have been present for days to years, with different factors likely to be contributory depending on duration and other factors.

Often there is limited range of motion and crackling. An increased cervical lordosis (chin forward) posture is often noted.

Especially look for trigger points on examination.

See further discussion at the beginning of the chapter.

Pathophysiology

Various factors may be causing pain and stiffness alone or in combination:

muscle tension due to stress, anxiety, or fatigue

incorrect posture causing strain on soft tissues and muscle spasm

nerve root irritation from postural factors, degenerative changes, or muscle spasm; this is especially likely to be an important factor when pain is unilateral and fairly acute

degenerative disc and joint changes lead to pain through various mechanisms: stresses on pain-sensitive facet joints, on pain-sensitive soft-tissue structures, and by nerve root irritation. This is likely to be a significant factor in older patients and in those with chronic pain and stiffness. An X-ray will show the extent of degenerative change

No matter what the inciting cause or causes (one or more of the above and/or an *acute strain*), all of these various factors can then interact and form vicious cycles:

pain → psychological tension → muscle tightness.

pain → muscle tightness directly.

pain → splinting → incorrect posture.

muscle tightness → incorrect posture → nerve root irritation.

degenerative changes → nerve root irritation.

nerve root irritation → muscle spasm and so on.

Prolonged limitation of motion leads to *loss of flexibility,* and therefore tissues which previously could tolerate certain positions and movements can no longer do so without pain.

Pain may also be due to "reflex myalgia" from TMJ or shoulder pain or

NECK AND UPPER BACK PAIN

headaches such as migraine; the primary source of the difficulty must obviously be addressed in these cases.

Treatment

General Measures

 Mild *analgesics* and *antiinflammatories* are helpful to break the pain cycle as well as for short-term relief.

 A *soft cervical collar* is useful in an acute episode or on an intermittent basis, but remember that prolonged use leads to loss of flexibility and muscle strength and may compound the problem.

 Massage with *cold* or the application of *heat* can give some relief.

Specific Measures (to be applied depending on history and findings; *any combination* may apply):

 In *chronic cases* or with obvious *postural problems*, patient education about proper ergonomics, posture and corrective exercise is paramount. A *cervical pillow* can be especially helpful if there is morning pain and stiffness. See Patient Handouts 4-2 and 4-3; strongly consider referral to a physical therapist.

 If trigger points are present, an effective treatment is *injection* of those points. See Figure 4-2.

 If *nerve root irritation* is a significant factor (especially likely when the problem is unilateral and fairly acute), antiinflammatory medication should be used in addition to intermittent *heat* and a *soft collar*. If resolution does not occur, *a trial of cervical traction* supervised by a physical therapist may be worthwhile. Traction is of course contraindicated in the presence of fracture, dislocation, rheumatoid arthritis or decreased bony strength (as from osteoporosis or metastatic disease).

 If *psychogenic muscle tension* contributes and the problem is acute, a short course of an antianxiety agent may help. In chronic cases these are contraindicated because of the potential for addiction, and the patient should be encouraged to take up an appropriate aerobic exercise program, to get help with stress reduction and/or use a *muscle relaxation technique* (meditation, self-hypnosis, or the method discussed in Patient Handout 4-5).

 In prolonged cases in which some *flexibility* has been *lost,* recommend the *mobilization exercises* shown in Patient Handout 4-4. These are contraindicated in patients with rheumatoid arthritis, acute injury, or decreased strength of bone.

 If there is *loss of muscle tone, strengthening exercises* supervised by a physical therapist should be considered. These are contraindicated in patients with intracranial vascular disease or significant uncontrolled hypertension.

 Because of the reinforcement and close follow-up that patients with chronic neck pain often need, it may sometimes be advantageous to refer them to a physical therapist or physiatrist. (See Table 4-1 for summary.)

36   4.   THE NECK AND UPPER BACK

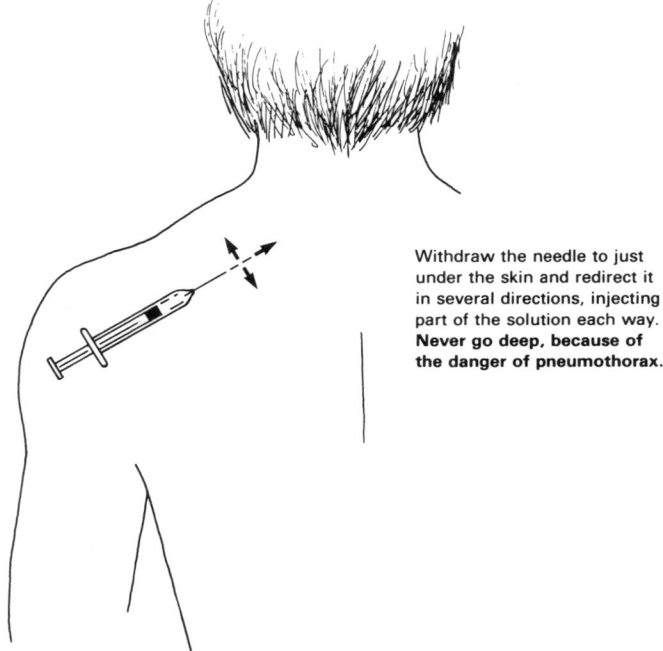

Fig. 4-2. Injection of trigger points. Find the tender nodule of muscle. Then prep the area and spray with ethyl chloride. Using a *short 25-gauge needle*, inject into the muscle at the point of maximum tenderness at an oblique angle and not too deep. About one cc of 1% xylocaine without epinephrine is about right; some advocate mixing in some steroid, while others feel that even the xylocaine can be dispensed with, and that saline does just as well by physically breaking up the tight knot of muscle. See pp. 23–27 for further precautions and details before injecting.

    A trial of *cervical traction* may be considered in chronic or recurrent cases. Traction is of course contraindicated in the presence of fracture, dislocation, rheumatoid arthritis, or decreased bony strength (as from osteoporosis or metastatic disease).
    Consider the possibility of a masked *depression* or fibrositis as the etiology of symptoms that are prolonged and not easily explained by the other factors discussed.
    Remember also the possibility of myalgia secondary to shoulder or TMJ pain, vascular headache, or rarely, serious intracranial or intrathoracic pathology or cervical neoplasm.

NECK AND UPPER BACK TRAUMA

Table 4-1   Therapeutic Measures in Neck Pain Syndrome

General: analgesic/antiinflammatories
ice/heat
instruction in posture/ergonomics (Patient Handout 4-3)
soft collar (short-term use only)

Specific Factor	Therapeutic Measure
morning pain/stiffness	cervical pillow
loss of flexibility	ROM exercise (Patient Handout 4-4)
loss of strength	exercise (P. T. referral)
trigger point	injection
nerve root irritation	soft collar short-term
	traction prn (P. T. referral)
psychogenic muscle tension	counseling, relaxation training, stress reduction, anxiolytic medication
depression	as appropriate
shoulder or TMJ pain	investigate and treat source
vascular/other headache	investigate and treat source
acute symptoms	soft collar short-term
prolonged symptoms	try traction; again rule out other causes
chronic symptoms	consider depression, chronic pain, physiatry referral

NECK AND UPPER BACK TRAUMA

Acute Strain of Specific Neck or Upper Back Areas
(Very Common)

Diagnosis

There is a history of acute strain, either with a sudden twist or prolonged abnormal posture.

Examination reveals tenderness localized in one area.

Always X-ray neck trauma.

Treatment

Advise rest of the involved area (temporary use of a soft collar may be useful when neck muscles are involved).

Heat and *analgesics* should be prescribed.

Cervical Sprain (Whiplash) *(Common)*

Note: If the patient presents acutely, the neck must be immobilized in neutral position (with sandbags if necessary) while adequate X-rays and neurologic examination are performed to rule out a fracture, dislocation, or spinal cord injury (any of which require continued immobilization and immediate neurosurgical consultation).

Diagnosis

Usually there is a history of forced flexion/extension of the neck.

The patient will commonly report that the onset of pain was not immediate after injury; often it begins the next day.

Diffuse tenderness is found.

The X-ray is normal.

If there are any neurologic findings, the patient may have a ruptured cervical disc or a spinal cord injury, and should be referred to an orthopedist or neurosurgeon.

Treatment

Use a *soft collar* acutely (with progressively less use as time goes on to prevent stiffness).

Even *bedrest* may be necessary for a short time in severe cases.

Ice application is useful in the first 48 hours; after that *wet heat* may be more beneficial.

Sufficient *analgesics* should be prescribed; *antiinflammatory drugs* may also be helpful.

As time goes on the various *other factors* discussed in the section on NECK PAIN SYNDROME may become operative and should be addressed as discussed there (for example: psychogenic muscle tension, loss of flexibility, loss of muscle strength).

Referral to a physical therapist or physiatrist for follow-up should be considered in cases that are at all severe or prolonged.

Be aware that *secondary gain* issues (financial and otherwise) are often involved in these cases.

PATIENT HANDOUT 4-1

Your Neck: Self Care

Some things you can do for your neck when it acts up:

1. Put *ice* packs on for twenty minutes or so at a time, or have someone massage the sore area with ice (allowing some water to freeze in a paper cup is an easy way to do this).
2. Put *wet heat* on, or stand under the hot shower.
3. Sometimes alternating the cold with the heat, one after the other, feels especially good.
4. Take a couple of aspirin or acetaminophen tablets to reduce inflammation and break the pain/spasm cycle. (Follow label directions.)
5. If stress is part of the problem, try to get away and relax a bit. Your doctor may give you more information or help with this.
6. Pay attention to what you've been shown about posture and exercise.

PATIENT HANDOUT 4-2

Using a Cervical Collar

A collar can be extremely helpful in the treatment of various neck problems, but only if it is worn properly. Be sure to follow these guidelines:

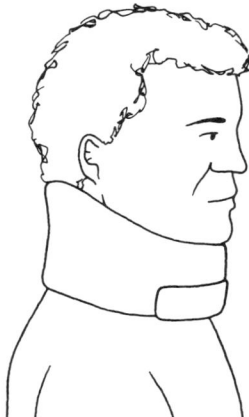

1. Make certain that it is the proper size to hold your head straight as in the picture, not tipped back or forward.
2. If you have been given the collar because of a recent strain or wryneck, leave it on for most or all of the day at first; after a couple of days begin wearing it less and less each day.
3. If your neck problem is long term, be sure to follow your doctor's advice as to how often and for how long to wear your collar.

PATIENT HANDOUT 4-3

Your Neck: Posture

Most people put extra strain on their necks because of poor posture. Try to be aware of the following:

1. Keep your chin tucked back.

2. Don't let your shoulders slouch forward.
3. When sitting, be sure you maintain an arch in your low back; a small pillow or rolled-up towel there can help you do this. And don't allow a high-backed chair or headrest to force your neck forward.
4. Avoid tilting your head backward, forward or to one side for a prolonged period, and don't hold your neck in one position for a long time. Move it around every so often.
5. Try not to sleep on your stomach, because your head will have to be twisted to the side. Either sleep on your side with your pillow just thick enough to hold your head straight, or if you sleep on your back use a "butterfly" pillow.* Special "cervical pillows" are available and often quite helpful.

Keep these general rules in mind when doing things like driving in your car, watching t.v., reading, talking on the telephone, shaving or putting on make-up, carrying a purse on one shoulder or a child in one arm, and when wearing bifocals.

---

*You can make one by tying a ribbon around the center of a soft pillow.

PATIENT HANDOUT 4-4

Relaxation

When you are tense, your muscles become tense, especially those in your neck, head, and back. After muscles have been tight for a while, they begin to hurt. If muscles are already sore from bad posture, a pinched nerve or whatever, those muscles will be the first to tighten up when you get tense. So one of the most important things you can do to help pain coming from muscles in your neck, back or head is to learn to relax.

There are a variety of techniques to do this. Some are called "meditation," some are called "self-hypnosis" and others have no specific name. You may wish to get a good book on one of these topics, such as *The Relaxation Response* by Herbert Benson, M.D. Taking up regular aerobic exercise such as walking, running, cycling, or swimming can also be extremely helpful. (Get your doctor's approval before starting.)

One relaxation technique is printed below. You may go through it at least twice a day to start; hopefully you will eventually incorporate relaxation into your approach toward life, or at least be able to grab hold of yourself and relax when stresses begin to pile up.

1. Lie down on your back, or sit in a comfortable chair. Close your eyes.
2. Breathe in slowly for a count of two and then out for a count of two. Keep breathing in this slow rhythm.
3. Make a fist in both hands and then relax them. Feel the relaxation spread up your arms and into your neck, and then down to your back and legs. Keep up the slow breathing.
4. Make a conscious effort to exclude from your mind anything but an awareness of your slow, rhythmic breathing, and thinking how very, very relaxed you are. Think about each part of your body in turn, and how relaxed it is; then think about your breathing; then think about each part of your body and how relaxed it is again.
5. Keep this up for at least 10 minutes.

Neck Exercises

Chin Tucks

Tuck in your chin without bending your neck; feel the back of your neck flatten. Hold for a few seconds and relax. Repeat ten times.

Shoulder Shrugs

With your arms relaxed at your sides, bring your shoulders up and then back; hold for a few seconds and then relax. Repeat ten times.

Range of Motion

With your chin tucked in as in the first exercise, turn your head as far as it will go to each side and hold for a few seconds; then lean your head to each side and hold for a few seconds. Still keeping your chin tucked in, do the same going forward. Repeat the whole process three times.

5

Neurologic Symptoms and Diffuse Pain of the Upper Extremity

Evaluation 44
    History. 45
    Examination 48
    Electrodiagnosis 51
    X-Ray 51
Syndromes. 51
    Cervical Radiculitis. 51
    Thoracic Outlet Syndromes. 52
    Brachial Plexitis 53
    Carpal Tunnel Syndrome 53
    Pronator Teres Syndrome 55
    Tardy Ulnar Palsy 55
    Ulnar Nerve Compression at the Wrist 56
    Radial Nerve Palsy 56

Further details on some therapeutic suggestions can be found in Chapter 3.

EVALUATION

    Pain and neurologic symptoms of the upper extremity can be due to *vascular compromise* by emboli, arteriosclerosis, Raynaud's phenomenon, etc. The

EVALUATION 45

discussion of these conditions is beyond the scope of this book, but their possibility must not be forgotten; appropriate history must be taken and necessary examination and tests performed in order to rule them out.

Also, weakness, numbness, or paresthesias obviously are often due to lesions of the *brain* or *spinal cord*. A *polyneuropathy* will usually manifest itself as sensory loss in a "glove" distribution without associated pain, but metabolic disease (especially *diabetes*) can affect single large nerves and lead to confusion with the compression neuropathies discussed herein. And diseases of *muscle* or the *neuromuscular junction* can underlie motor weakness.

Again, discussion of these diseases is beyond the scope of this volume; but it is imperative that further history, examination, and testing—often well beyond that which is discussed below—be undertaken to rule out such an etiology whenever the original history and/or examination is at all suggestive. These diseases usually will not have significant pain associated with the neurologic symptoms or findings, but this is not a completely reliable clue. And do not forget that an arm ache can be a manifestation (sometimes the only one) of serious *intrathoracic disease* such as angina pectoris, an acute myocardial infarction or an apical pulmonary tumor. Index of suspicion must remain high, and threshold for doing ECG, chest X-ray etc. kept low.

History

Ask about the *duration* of symptoms and the *location* of *pain* and *paresthesias*. Often the answer to the latter question is too vague to be useful. See Figure 5-1. A history of nonspecific aching in both shoulders and upper arms in an older patient without other specific history or findings is suggestive of POLYMYALGIA RHEUMATICA, discussed in Chapter 21.

Has the patient noticed any *weakness* or *incoordination*?
Inquire about precipitating factors:
Pain mainly at night, especially after much wrist usage during the day, is suggestive of the CARPAL TUNNEL SYNDROME.
Paresthesias and pain which come on with the arm in certain positions (especially while sleeping or with the arm leaning on a hard object) are suggestive of a THORACIC OUTLET SYNDROME or peripheral nerve compression.
Pain brought on by general arm usage and relieved with rest should bring to mind the possibility of vascular disease.
Continuous symptoms, especially if pain is not a significant factor, must lead to consideration of a CNS lesion.
Bilateral symptoms, which may or may not also involve the lower extremities, should immediately raise suspicion of a cervical cord compression syndrome, either due to a cervical disc protrusion, a fracture, or sometimes neoplasm. (With a disc protrusion neck flexion usually exacerbates the

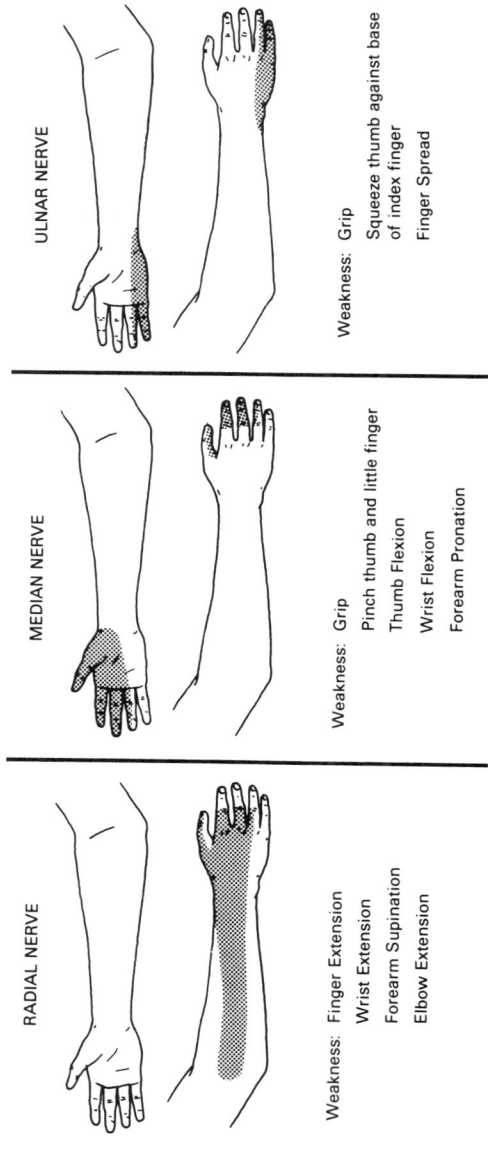

Fig. 5-1. Common root and peripheral nerve sindromes (upper extremity). Syndromes are usually incomplete. Dermatomes and innervations shown can vary in some patients.

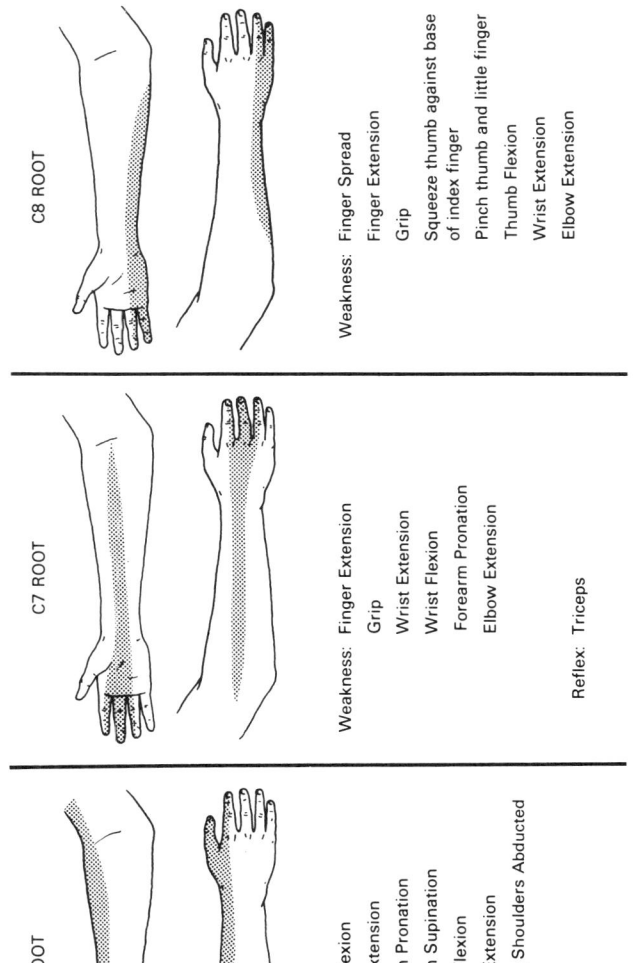

Fig. 5-1. Continued.

symptoms.) Immediate cervical immobilization and neurosurgical consultation is imperative.

Pain brought on by exposure to cold, especially if associated with color changes, raises the possibility of *Raynaud's Phenomenon*.

Intermittent numbness and tingling in both arms, especially if brought on by stress or associated with perioral numbness, is suggestive of *hyperventilation*. See if the symptoms can be reproduced with rapid deep breathing in the office.

Has there been any *neck pain* or recent *neck trauma*? These would suggest CERVICAL RADICULITIS.

Question the patient about any *weakness, numbness, paresthesias* or *incoordination* elsewhere in the body, or any headaches, visual or auditory symptoms, vertigo, imbalance, speech difficulty, episodes of loss of consciousness, or recent head or neck trauma, any of which may suggest a CNS lesion.

Examination

Examine the *shoulder, elbow,* and *wrist* for *range of motion, swelling,* and *tenderness,* since often a specific articular or periarticular syndrome (e.g., shoulder bursitis, epicondylitis) will be the cause of the patient's complaints even though he or she is unable to localize the symptoms.

Examine the *neck* and paracervical area for tenderness and check range of motion of the cervical spine.

Check distal *pulses* and *capillary filling*.

Elicit the biceps or brachioradialis and the triceps *reflexes* (Figure 5-2).

Test muscle *strength*. See Table 5-1 for the various motions to be tested and their root and peripheral innervations. Significant weakness is always an indication for referral for more precise diagnosis (e.g., EMG) and more aggressive management.

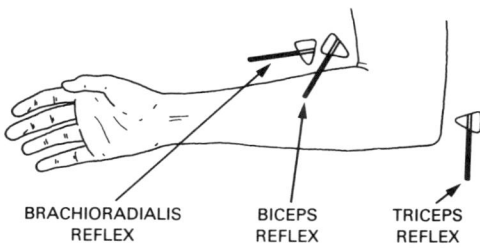

BRACHIORADIALIS REFLEX   BICEPS REFLEX   TRICEPS REFLEX

Fig. 5-2. Reflexes: upper extremity.

EVALUATION

Table 5-1   Outline for Examination of Patients with Diffuse Pain and/or Neurologic Symptoms in the Upper Extremity

1. Check for tenderness around, and range of motion of, the wrist, elbow and shoulder
2. Check for tenderness and range of motion of the neck
3. Check vascular status: radial and ulnar pulses, distal skin color and temperature, and capillary filling (press the nail and see how long it takes the color to return to the nailbed)
4. Check the biceps or brachioradialis (C5-6) and the triceps (C6-7) reflexes
5. Test muscle strength by comparing with the uninvolved side (all maneuvers to be done against your resistance; root and peripheral innervations are as noted below)
   FINGER SPREAD (C8, T1; ulnar)
   HOLDING FINGERS EXTENDED AT MCP AND PIP JOINTS (C7,8; radial;)
   GRIP (RADIAL AND ULNAR SIDES) (C7-T1; median and ulnar)
   SQUEEZE THUMB AGAINST BASE OF INDEX FINGER (C8,T1 ; ulnar)
   PINCH BETWEEN THUMB AND LITTLE FINGER (C8, T1; median)
   FLEX THUMB (C8; median)
   EXTEND WRIST (C6–8; radial)
   FLEX WRIST (C6–7; median)
   SUPINATE FOREARM WITH ELBOW STARIGHT (C5–6; radial)
   PRONATE FOREARM WITH ELBOW STRAIGHT (C6–7; median)
   FLEX ELBOW WITH FOREARM SUPINATED (C5–6; musculocutaneous)
   STRAIGHTEN ELBOW (C6–8; radial)
   HOLD SHOULDERS ABDUCTED (C5–6; axillary)
6. Screen pin sensation in the following areas (innervations shown):
   WEB SPACE BETWEEN THUMB AND INDEX FINGER ON THE BACK OF THE HAND (C6; radial)
   PALMAR SIDE OF THE DIP JOINT OF THE MIDDLE FINGER (C7; median)
   SIDE OF THE HAND PROXIMAL TO THE FIFTH DIGIT (C8; ulnar)
   SIDE OF THE MID-FOREARM PROXIMAL TO THE FIFTH DIGIT (T1)
7. Perform the specific maneuvers shown in Figure 5-3 in sequence.

[a]With some practice the entire examination will take no more than a few minutes. Of course, additional examination should be done as indicated.

Look for muscle *atrophy* (especially that seen in the thenar eminence with severe or prolonged CARPAL TUNNEL SYNDROME).

Test for pin *sensation*. When a more sensitive examination needs to be done, check two-point discrimination and compare closely to the uninvolved side. See Figure 5-1 for the areas classically involved with the various cervical roots and peripheral nerves; but note that areas are often incompletely involved, that sensation testing is subjective and thus quite often inaccurate unless the deficit is almost complete, and that there are individual variations from the dermatomes shown.

Do the following maneuvers (Figure 5-3); reproduction or intensification of symptoms help to define the problem as listed:

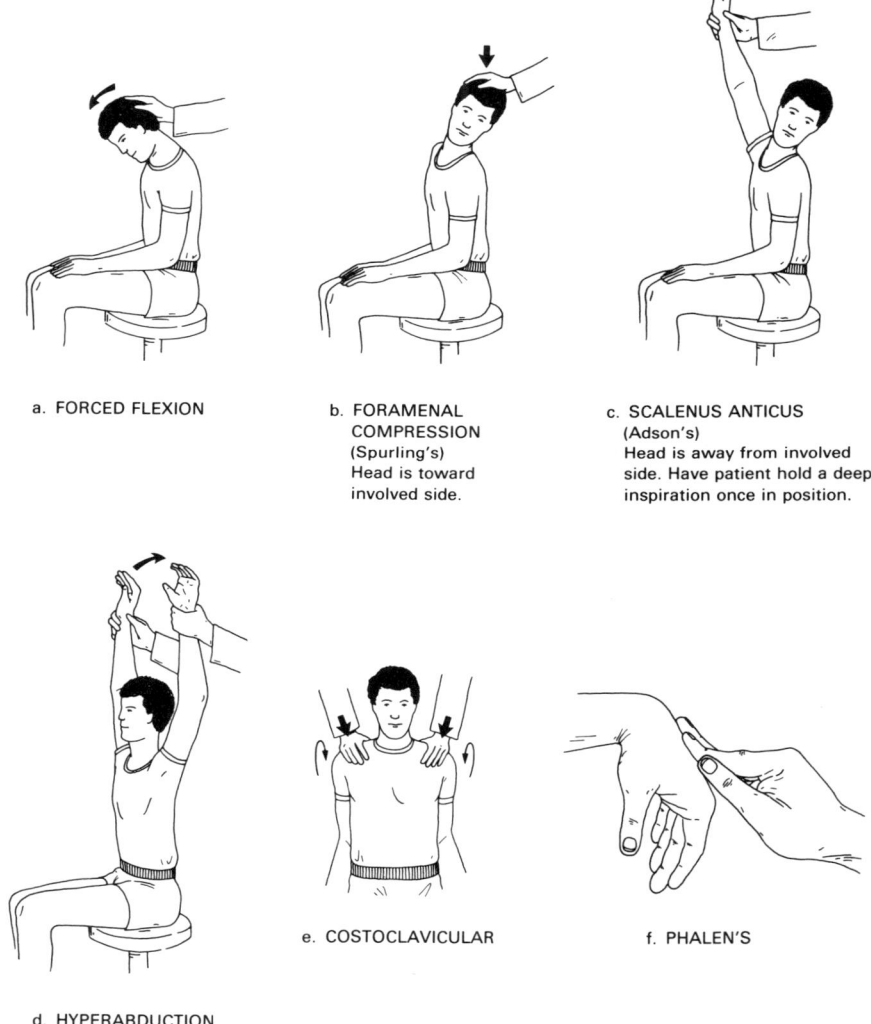

a. FORCED FLEXION

b. FORAMENAL COMPRESSION (Spurling's) Head is toward involved side.

c. SCALENUS ANTICUS (Adson's) Head is away from involved side. Have patient hold a deep inspiration once in position.

d. HYPERABDUCTION (Wright's)

e. COSTOCLAVICULAR

f. PHALEN'S

Fig. 5-3. Maneuvers. Each should be maintained for at least 10 seconds, except Phalen's maneuver, which should be held for 30 seconds. Reproduction of distal symptoms is a positive result.

SYNDROMES

Full flexion of the neck: CERVICAL RADICULITIS on a discogenic basis
The foramenal compression test (Spurling's maneuver): CERVICAL RADICULITIS from nerve root impingement in the neural foramen. Never manipulate the neck until fracture has been ruled out.
The thoracic outlet maneuvers:
Adson's maneuver: the ANTERIOR SCALENE SYNDROME
Costoclavicular maneuver: the COSTOCLAVICULAR SYNDROME
Hyperabduction maneuver: the PECTORALIS MINOR SYNDROME
Phalen's maneuver: the CARPAL TUNNEL SYNDROME

Electrodiagnosis

When the diagnosis cannot be pinpointed by the history and examination described above, referral should be made for an EMG/NCV (described in Chapter 2).

X-Ray

A cervical spine X-ray and a chest X-ray should be obtained in every patient with objective neurologic findings not clearly due to a peripheral problem such as CARPAL TUNNEL SYNDROME. Referral for further evaluation is warranted in cases with severe or progressive deficit or if there is suspicion of a spinal cord lesion.

SYNDROMES

Cervical Radiculitis ("Pinched Nerve") *(Very Common)*

Diagnosis

Almost invariably there is pain and tenderness in the neck as well as distal symptoms.

If pain, paresthesias, and numbness can be localized, then the pattern is consistent with one or more cervical dermatomes.

Any muscle weakness and hyporeflexia is consistent with the involved roots. (Remember that significant or progressive weakness is an indication for referral for further investigation to rule out tumor, etc.) See Figure 5-1.

Neck flexion (Figure 5-3a) reproduces or aggravates the distal symptoms when root irritation is by a disc.

The foramenal compression test (Figure 5-3b) causes distal symptoms if degenerative arthritis with foramenal encroachment is responsible; this is much more common.

An X-ray should be done, and often will show evidence of degenerative arthritis, foramenal encroachment, and/or disc disease at the involved levels.

5. NEUROLOGIC SYMPTOMS AND DIFFUSE PAIN OF THE UPPER EXTREMITY

Pathophysiology

Impingement on one or more nerve roots is by a bulging or herniated disc, or by narrowing of the exit foramena, or from degenerative hypertrophy of the posterior joints.

Inflammation and swelling around the root may be an initiating or a contributing factor as the root travels in a previously narrowed but not yet impinging pathway.

Treatment

Use an *antiinflammatory medication*.

A *soft cervical collar* should be worn, with its use tapered as the patient responds (Patient Handout 4-2).

Wet *heat* to the neck may be helpful.

If these measures do not bring improvement, a trial of *cervical traction* should be considered. Traction is contraindicated in rheumatoid arthritis, acute trauma, instability, and conditions causing decreased strength of bone.

Cases with significant or progressive motor weakness, or which are not responding to these measures, should be referred to the specialist.

Thoracic Outlet Syndromes *(Common)*

Diagnosis

Diffuse pain and paresthesias are usually positional.

Symptoms are reproduced by Adson's maneuver (SCALENUS ANTICUS SYNDROME), the costoclavicular maneuver (the COSTOCLAVICULAR SYNDROME), or by the hyperabduction maneuver (the PECTORALIS MINOR SYNDROME) (Figure 5-3).

X-ray may show a cervical rib, but this is probably not often the cause of the patient's symptoms. Chest X-ray should be done to rule out an apical chest lesion.

Pathophysiology

Compression of the neurovascular bundle occurs as it passes through and around various structures: between the scalenus anticus muscle and the first rib, between the first rib and the clavicle, or between the pectoralis minor muscle and the first rib.

Stress, poor posture, and muscular fatigue are usually what allow narrowing at these various locations.

Treatment

In patients with intermittent symptoms, *avoidance* of the responsible position is all that is necessary.

Patients with more persistent symptoms should be advised to practice consciously not allowing the shoulders to droop and to follow the precautions itemized in Patient Handout 4-3.

Surgery to remove a cervical rib is rarely necessary.

Brachial Plexitis *(Rare)*

Diagnosis

The patient presents with fairly acute, diffuse, and usually severe axillary, arm and hand pain, paresthesias and often weakness, often without known trauma.

Especially look for proximal muscle atrophy and scapular winging.

There is usually axillary tenderness on examination.

Cervical spine and chest X-rays must be done to rule out lesions there.

Blood sugars should be done to rule out a diabetic neuropathy.

Pathophysiology

Idiopathic inflammation of the brachial plexus is responsible, possibly on a postviral autoimmune basis.

Treatment

Place the arm in a *sling*.

Potent *antiinflammatory medication* should be used.

The problem sometimes resolves in a week or two. Referral to a neurologist is wise if the patient does not improve fairly quickly.

Carpal Tunnel Syndrome (Median Nerve Compression in the Wrist) *(Common)*

Diagnosis

The patient complains of aching pain in the first three digits and radial side of the hand, often with paresthesias and numbness. Note that the pain is often more diffuse and can extend well up the arm.

Symptoms are usually predominant at night and are more pronounced following days in which the wrist has been overused (common in carpenters and supermarket checkers, for example).

On examination there may be decreased sensation in the median nerve distribution (Figure 5-1).

Phalen's maneuver should produce aching pain and paresthesias (Figure 5-3f).

Tapping the middle of the volar wrist may more easily produce distal paresthesia than in the uninvolved extremity (Tinel's sign).

Check for atrophy of the thenar eminence or objective weakness (best evaluated by testing the strength of opposition of the tip of the thumb to the tip of the little finger, and of extension of the second and third digits at the PIP joint with the MCP joint hyperextended).

The condition is common in pregnancy, and also found in diabetes mellitus, acromegaly, rheumatoid arthritis, and myxedema. The latter four conditions should be tested for in severe, prolonged or otherwise unexplained cases, and X-rays to rule out impinging bony lesions should be ordered as well.

Pathophysiology

Swelling in the carpal tunnel of the wrist causes compression of the median nerve. Rarely, a bony lesion or soft-tissue mass is responsible for the compression.

Treatment

A cock-up *wrist splint* (Figure 8-3) should be used during the day and especially during offending activity, even if most of the symptomatology is nocturnal.

Oral *antiinflammatory medications* should be prescribed.

If these measures are unsuccessful, *injection* of steroid should be offered. Success rate is good. See Figure 5-4.

If that is unsuccessful, or if at any time there is found thenar atrophy or motor weakness on examination, the patient should be referred for confirmatory NCV testing and then *surgery* (in which the transverse carpal ligament is incised).

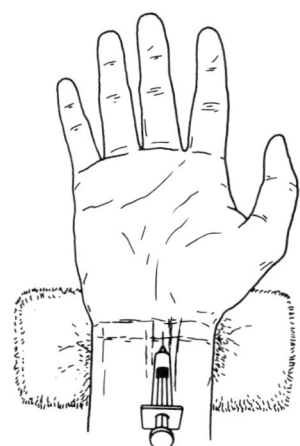

Fig. 5-4. Carpal Tunnel Injection. have the patient hold his wrist dorsiflexed over the edge of the treatment table or over a rolled towel. Find the distal transverse crease and inject just ulnar to the prominent palmaris longus tendon, with the needle pointing at a 45° angle toward the hand. Use a 25-gauge needle and inject about a ½ cc. of steroid and ½ cc. of xylocaine without epinephrine. See p. 24 for further details and precautions before injecting.

Pronator Teres Syndrome (Median Nerve Compression in the Forearm) *(Rare)*

Diagnosis

Usually a history is given of direct trauma or compression of the proximal forearm or excessive pronation/supination movement.

Symptoms and neurologic exam are analogous to the CARPAL TUNNEL SYNDROME. (Differences are usually too subtle to detect.)

Phalen's test is negative.

Full pronation of the forearm intensifies the symptoms.

There is often tenderness to deep palpation of the proximal volar forearm.

Pathophysiology

The median nerve is compressed as it passes between the two heads of the pronator teres muscle.

Treatment

Have the patient *avoid* the precipitating compression or activity.

If weakness or atrophy develops, referral should be made for electrodiagnostic confirmation and *surgery*.

Tardy Ulnar Palsy (Cubital Tunnel Syndrome; Ulnar Neuritis; Ulnar Nerve Compression at the Elbow) *(Uncommon)*

Diagnosis

Found in patients who lean on their elbow or who have traumatized their elbow and sometimes in baseball pitchers.

Paresthesias and possibly decreased sensation occur in the fourth and fifth digits.

Pressure on the cubital tunnel (between the olecranon and the medial epicondyle) reproduces or intensifies the symptoms.

When severe, weakness of finger spread and interosseus atrophy can be found.

Pathophysiology

The ulnar nerve suffers acute or repetitive injury in the cubital tunnel.

Treatment

Use *padding* and/or *avoidance* of the responsible position.

Refer for electrodiagnostic confirmation and *surgery* when symptoms are prolonged or severe, or if there is motor weakness or atrophy.

Ulnar Nerve Compression at the Wrist *(Uncommon)*

Diagnosis

Symptoms and neurologic findings are as in a TARDY ULNAR PALSY. Palpation of the pisiform-hamate tunnel may cause symptoms.

Pathophysiology

Compression of the ulnar nerve can occur where it passes through the "canal of Guyon" between the pisiform and hamate bones or can be secondary to trauma or to bony or soft tissue anomolies.

Treatment

If there are significant sensory symptoms or any motor findings, the patient should be referred for electrodiagnostic confirmation and *surgical release*.

Radial Nerve Palsy *(Rare)*

Diagnosis

An acute radial nerve palsy will be obvious (wrist drop, imparied sensation) and follows prolonged compression, usually during unconsciousness.

More chronic symptoms can be produced by compression of the nerve as it passes in the area around the lateral epicondyle.

Treatment

Referral for management by the specialist is appropriate.

6

The Shoulder

Shoulder Pain and Limitation of Motion 57
    History 60
    Examination 60
    Rotator Cuff Tendonitis and Subacromial Bursitis 64
    The Process of Shoulder Pain 65
    Biceps Tendonitis 66
    Acromioclavicular Joint Arthritis 67
    Adhesive Capsulitis 68
    Rotator Cuff Tear 68
    Degenerative Arthritis of the Shoulder 69
    Shoulder Instability 70
    Recurrent Subluxation 70
Shoulder Trauma 70
    Acromioclavicular Joint Sprain 71
    Clavicle Fracture 71
Shoulder Strain 71

Pain in the *neck muscles* is discussed in Chapter 4. Further details on therapeutic suggestions made in this chapter can be found in Chapter 3.
    For the anatomy of the shoulder, see Figure 6-1.

SHOULDER PAIN AND LIMITATION OF MOTION

Note: Always consider the possibility of shoulder pain being *referred* from *intrathoracic disease* or diaphragmatic irritation (ischemic heart disease, pulmonary tumors, and gall bladder disease, for example). While the differential diagnosis and management of such conditions is obviously beyond the scope of this

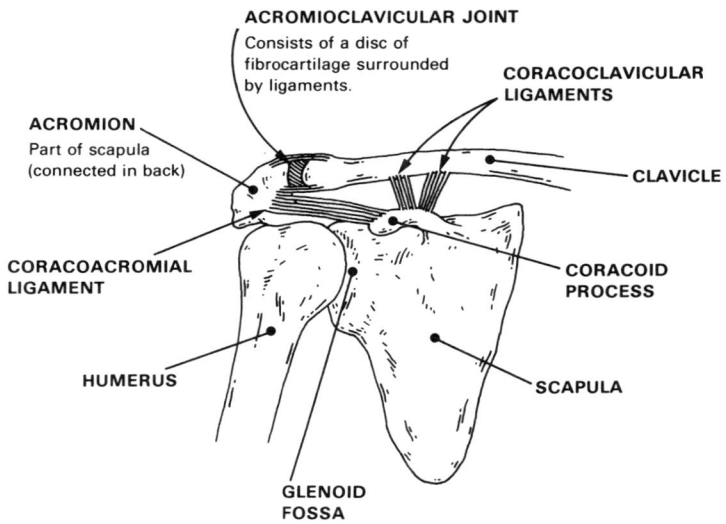

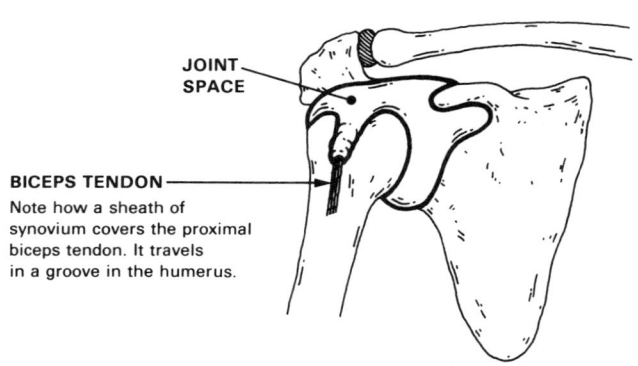

Fig. 6-1. Anatomy of the shoulder.

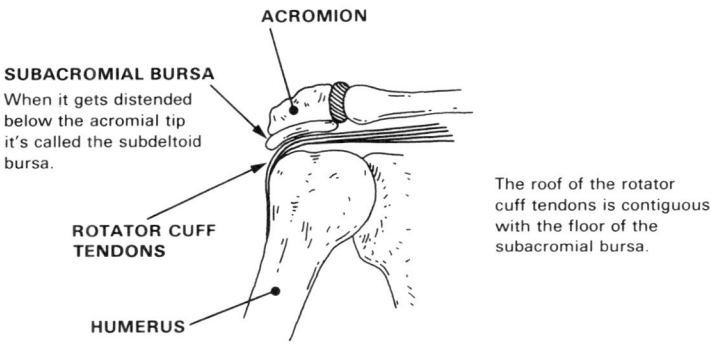

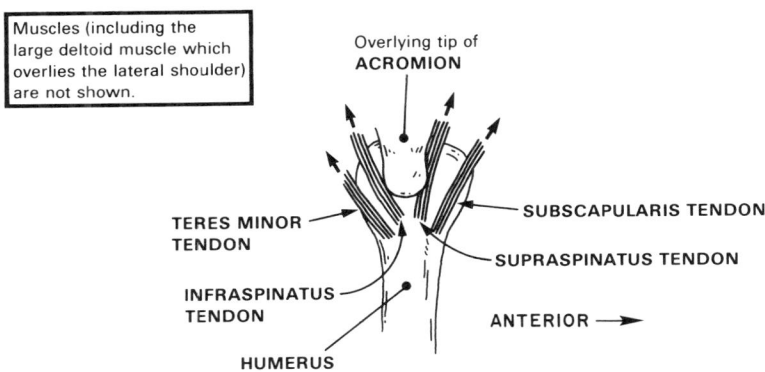

Fig. 6-1. Continued

book, they must be kept in mind and appropriate history, examination, X-rays and other tests done to rule them out, especially if the initial history and/or examination is not completely consistent with a musculoskeletal origin of the patient's pain. And don't overlook the possibility of bone tumor or metastasis, occult trauma or aseptic necrosis presenting with shoulder pain.

History

Ask about the *duration* and *location* of the pain. (If the patient is actually referring to the posterior musculature rather than the shoulder area itself, refer to Chapter 4 on neck pain).

What are *precipitating* and *relieving* factors?

Has there been any specific *strain, overuse,* or *trauma*?

Inquire about any *neck pain* or *trauma, radiation* of pain into the arms or neurologic symptoms in the arms. (Again, if any are present see Chapter 4 on neck pain or the discussion of radiated arm pain in Chapter 5.)

Ask about *previous shoulder problems,* diagnoses, and treatments.

Are any *other joints* involved? Does the patient have a history of arthritis or gout?

Examination

Inspect for *deformity,* including a winged scapula (indicative of thoracic nerve palsy), muscle atrophy, etc.

Look for *tenderness* (see Figure 6-2).

Test *range of motion* (see Figure 6-3).

Limitation (due to pain) of elevation only, or elevation and one or two other motions, indicates ROTATOR CUFF TENDONITIS (but is also consistent with ACROMIOCLAVICULAR JOINT ARTHRITIS if tenderness was found there).

The impingement sign (Figure 6-4) is usually positive with rotator cuff problems.

Limitation of all motions implies ADHESIVE CAPSULITIS or much more rarely DEGENERATIVE ARTHRITIS OF THE SHOULDER.

A positive apprehension sign may indicate recurrent subluxation (which the patient may not have noted as such). (Figure 6-4.)

Look for *swelling, redness,* or *warmth* which would imply an acute inflammatory process.

Check *distal strength, sensation,* and *pulses.*

Test *strength* (see Figure 6-5). If true weakness (not just lack of effort related to pain) is present in certain motions, a ROTATOR CUFF TEAR is suggested, especially with prolonged symptoms or following trauma.

BICEPS TENDONITIS
Always compare with the uninvolved side, because this area is often a bit tender anyway. Best felt with the shoulder externally rotated. Confirm by producing pain by resisting the patient's attempt to flex the elbow, or to supinate the forearm with the elbow bent and against his side.

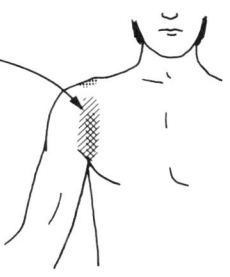

ANTERIOR

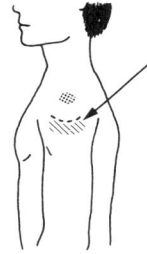

ROTATOR CUFF TENDONITIS/ SUBACROMIAL BURSITIS
There will be pain on passive elevation of the shoulder (past 45° to 110°), and possibly pain with certain (but usually not all) other motions as well, depending on the tendon or tendons involved.

LATERAL

ADHESIVE CAPSULITIS
There will be some restriction of motion in all directions.
OR
AN INTRAARTICULAR PROCESS
Consider *degenerative arthritis* in an older patient or one with previous trauma. If there is redness, swelling or warmth the joint must be tapped to rule out infection (Fig. 6-3).

AC JOINT ARTHRITIS
Extreme elevation will be painful.

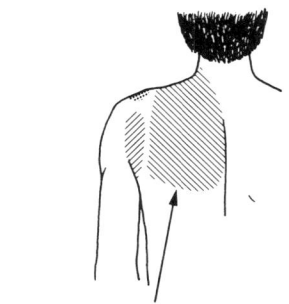

POSTERIOR MUSCULATURE
See chapter 4.

POSTERIOR

Fig. 6-2 Sites of tenderness: the shoulder.

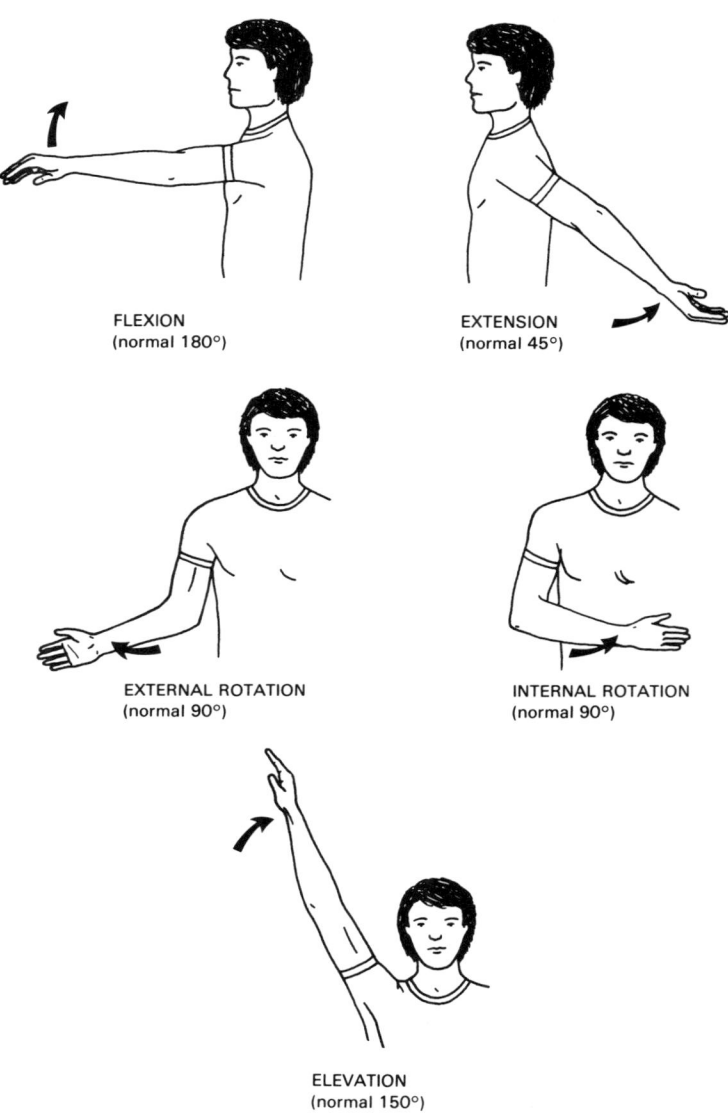

Fig. 6-3. Shoulder range of motion.

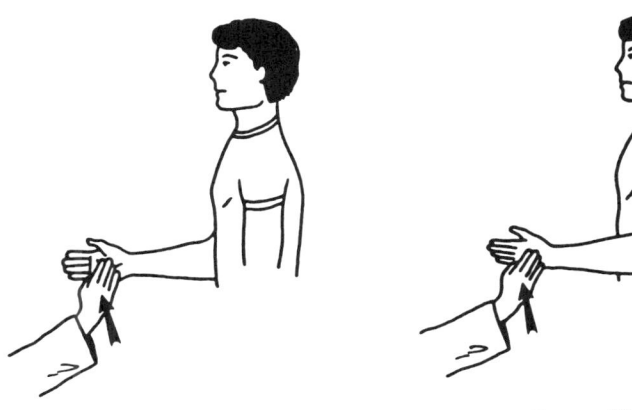

Internal Rotator
(subscapularis)

External Rotator
(infraspinatus, teres minor)

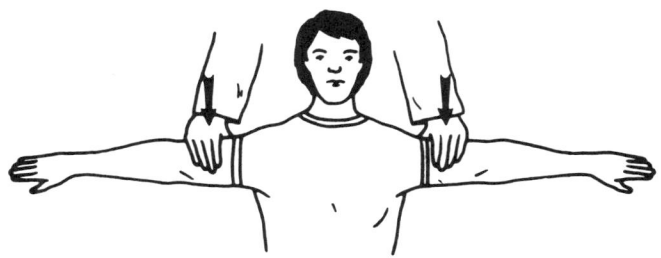

Supraspinatus

Deltoid and Supraspinatus

Fig. 6-4. Shoulder signs

6. THE SHOULDER

Impingement Sign

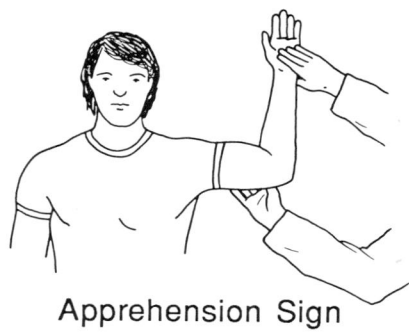

Apprehension Sign

Fig. 6-5. Checking rotator cuff strength. True weakness must be differentiated from the more common lack of effort due to pain.

Rotator Cuff Tendonitis and Subacromial Bursitis
(impingement syndrome, shoulder tendonitis, subdeltoid bursitis, calcific tendonitis) *(Extremely Common)*

Diagnosis

The symptoms may be very acute, very chronic or anywhere in between. (The so-called *hyperactive* shoulder is probably secondary to the release of calcium crystals into the bursa.)

A history of specific overuse (usually with the arm above neck level) is common but not invariable. Very common in swimmers, pitchers, house painters, etc.

The pain may be referred distally to the area of the deltoid muscle insertion.

You may find subacromial tenderness or no localizable tenderness. Tenderness of the paracervical muscles is often associated, and is probably due to a "reflex myalgia" secondary to the shoulder pain.

SHOULDER PAIN AND LIMITATION OF MOTION

Elevation is painful past a certain point (anywhere from 60° to 105°). Often other motions (but usually not all) are painful as well. True restriction of motion (not just because of pain) indicates that CAPSULITIS has developed, while true weakness in the setting of prolonged symptoms or following trauma suggests a ROTATOR CUFF TEAR.

The X-ray may reveal calcium in the bursa or along the tendon in long-standing cases.

See the box above.

The Process of Shoulder Pain

MOST SHOULDER PAIN is the result of a single pathologic process which, depending on how far it has proceeded and what paths it has taken, presents as various clinical syndromes and carries different names. Understand it, and you will understand shoulder pain.

It starts with the anatomic fact that the rotator cuff tendons, which stabilize the head of the humerus in the glenoid and aid in rotation and elevation of the shoulder, pass under a rigid structure called the "coracoacromial arch," made up of the bony acromion process and the coracoacromial ligament. When the shoulder is flexed and elevated (as in Figure 6-7) the tendons are compressed. Intermittent pain with this motion is termed the "impingement syndrome."

Now, this area of the rotator cuff happens to have a marginal blood supply to begin with, and so with repeated impingement is especially liable to become inflamed . . . thus "rotator cuff tendonitis."

The subacromial bursa is a potential sac which serves to lubricate between the rotator cuff and the coracoacromial arch. When the tendons are inflamed, so is the bursa; thus "subacromial bursitis."

With long-term impingement and inflammation, degenerative change tends to occur in the rotator cuff, often associated with calcium deposition: "calcific tendonitis."

Deterioration of the cuff can lead to microtears which can enlarge to "rotator cuff tears" and make symptoms hard to cure conservatively.

Inflammation in the cuff can easily spread to involve the biceps tendon, since that structure shares the synovial sheath with the cuff and itself can be impinged upon; this is the main cause of "biceps tendonitis."

Inflammation can also spread to the underside of the acromioclavicular joint and cause pain there.

And if it spreads to the joint capsule, that along with the lack of motion resulting from pain leads to "adhesive capsulitis."

Remember that the rotator cuff stabilizes the head of the humerus in the glenoid and insures proper motion. So deteriorating rotator cuff function can be a factor in the development of degenerative arthritis of the glenohumeral joint itself.

Treatment

Advise *rest* and *avoidance* of the responsible activity. Use a sling for a few days only with severe symptoms.

However, in cases that drag on for weeks, especially in older patients, it is imperative that *active range of motion* be maintained (see Patient Handout 6-1 for a series of exercises to prevent the development of ADHESIVE CAPSULITIS; do not have the patient do the sideways wall climbing).

Ice is helpful especially in acute cases. Some patients report better relief with heat.

Oral *antiinflammatories* may be used.

Steroid injection has a very good success rate and is considered by many to be the treatment of choice, especially in the hyperacute situation. See Figure 6-6.

In long-term cases, lack of response may be due to the presence of a partial ROTATOR CUFF TEAR, either degenerative in nature or secondary to unrecognized trauma. Referral is advised.

In the absence of a tear, *surgery* should be considered a last resort, though sometimes removal of calcium deposits or partial acromionectomy does give good results.

In swimmers, pitchers and other *athletes,* proper mechanics, balanced muscle strength and adequate flexibility are most critical. Refer to a physical therapist for the latter two, and advise the patient that obtaining advice from a good coach is most important to allow early return to activity and prevent recurrence.

Biceps Tendonitis *(Common)*

Diagnosis

Tenderness is found over the biceps tendon with the shoulder externally rotated. Always compare to the uninvolved side.

The diagnosis is confirmed by reproducing pain there with resisted elbow flexion and/or resisted forearm supination with the elbow bent.

Pathophysiology

Inflammation and degenerative changes of the tendon occur due to overuse, often in association with rotator cuff problems (see the box above). Sometimes the tendon pops out of its groove recurrently (the most common cause of the "snapping" or "popping" shoulder) leading to local irritation.

Treatment

Use *rest, ice,* and *antiinflammatories*.

Injection may be offered. Success rate is good. See Figure 6-7.

Note that degenerative changes in the tendon (and possibly steroid injection too) can predispose to *rupture* with the characteristic appearance of a bulge in the

SHOULDER PAIN AND LIMITATION OF MOTION

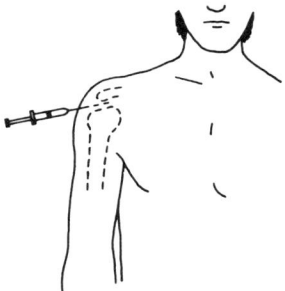

Fig. 6-6. Injection for cuff tendonitis and subacromial bursitis. Prep the area and spray with ethyl chloride. Use a 23-gauge needle. Inject a cc each of steroid and anesthetic subacromially as shown. The patient must be told to avoid ballistic motions (such as throwing a ball or swinging a racket) for 3 weeks or so. Injections may be repeated at weekly intervals if necessary up to a total of three in a course. See pp. 24-26 for further details and precautions.

distal upper arm. Repair is necessary if the patient requires powerful supination of his forearm at work or play, or for cosmetic reasons. If it is to be done, however, it must be done soon after rupture.

Acromioclavicular Joint Arthritis *(Uncommon)*

Diagnosis

Tenderness is found at the acromioclavicular joint (see Figure 6-2).
Pain occurs with high elevation of the shoulder.

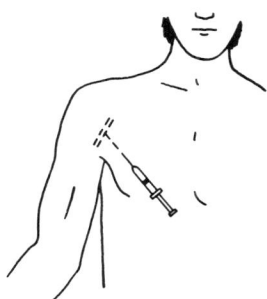

Fig. 6-7. Injection for bicepital tendonitis. Have the patient hold the arm externally rotated. Find the tender biceps tendon and after prepping the area and spraying with ethyl chloride, use a 23-gauge needle to inject a cc of steroid and a cc of local anesthetic around (never into) the sore tendon. The patient must be told to avoid ballistic motions for at least 3 or 4 weeks. See pp. 24-26 for further details and precautions.

Pathophysiology

Degenerative changes in the fibrocartilagenous disc that connects the acromion to the clavicle occur due to previous recognized or unrecognized trauma. Inflammation in the rotator cuff may also spread to this area.

Treatment

Use *rest, ice* or *heat*, and *antiinflammatories*.

Steroid injection may be tried but is quite painful to perform and success rate is only fair. The method is simply to inject into the tender joint from above. See Chapter 3 for further details and precautions.

Adhesive Capsulitis (Frozen Shoulder) *(Uncommon)*

Diagnosis

The patient presents with progressive diffuse pain and global limitation of motion of the shoulder, often but not always following another recognized cause of shoulder pain.

Pathophysiology

Fibrotic, restrictive changes in the tissues of the shoulder capsule occur secondary to disuse.

Treatment

Progressive *range of motion exercises* are essential. See Patient Handout 6-1. Include the sideways wall climb. Since these patients need close follow up to reinforce the program and check on their progress, *referral* to a *physical therapist* is wise. The therapist may also use mobilization techniques.

Oral *antiinflammatory medications* should be used.

It may be quite helpful to use a series of three weekly intraarticular and subacromial *steroid injections*. See Figures 6-6 and 6-8.

The condition usually improves in a few months with varying degrees of residual disability.

Referral for manipulation under anesthesia or surgery may be necessary if improvement is not adequate.

Rotator Cuff Tears *(Uncommon)*

Diagnosis

These are usually degenerative in nature and usually present as an especially prolonged or severe cuff tendonitis-like syndrome. They can present following specific trauma as well.

SHOULDER PAIN AND LIMITATION OF MOTION

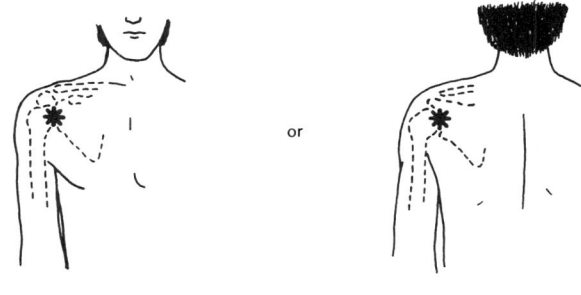

ANTERIOR **POSTERIOR**

Fig. 6-8. How to aspirate or inject a shoulder joint. Feel the head of the humerus and move your finger medially until it curves away from you and you feel the joint line. (You will be just lateral to the coracoid.) Alternatively, move medially on the posterior surface of the humeral head until the joint line is encountered. Prepare the site and use sterile technique throughout. After ethyl chloride spray, enter the joint perpendicularly using a 20-gauge or 18-gauge needle at least 1½ inches long, and aspirate. If steroid is to be injected, use a hemostat to hold the needle bevel and switch syringes to one containing 2 cc of steroid. (Never inject steroid if the synovial fluid is cloudy or if there is other reason to suspect infection.) Steroid can be injected the same way, without aspiration, using a 22-gauge needle. See pp. 14 and 24 for discussion of synovial fluid analysis and of precautions and risks *re* steroid instillation.

A big problem in baseball players.

On examination there is weakness or loss of active elevation past a certain point. The latter is sometimes difficult to differentiate from unwillingness to elevate the arm secondary to pain from tendonitis.

X-ray may reveal a high-riding humeral head.

Referral for anthrography or arthroscopy may be necessary to make the diagnosis.

Pathophysiology

See the box on p. 65.

Treatment

The patient should be referred to an orthopedist for consideration of surgery, except in cases of partial degenerative tears where the disability from the tear does not warrant such consideration.

Degenerative Arthritis (Osteoarthritis) of the Glenohumeral Joint *(Uncommon)*

Diagnosis

Chronic pain is worse after use.
Found in older patients or those with previous significant trauma.

There will be varible diffuse tenderness and variable pain with all directions of motion.

Pathophysiology

See the box on p. 65 and Chapter 22.

Treatment

See Chapter 22.
Technique of injection is shown in Figure 6-8.

SHOULDER INSTABILITY

Recurrent Subluxation *(Uncommon)*

Diagnosis

The patient reports a history of the shoulder recurrently and painfully feeling like it goes out of place. There is almost always a history of previous shoulder dislocation. Especially common in teenagers and young athletes. (Note that some loose-jointed people can pop their shoulder in and out, but this isn't associated with pain or previous dislocation.)

On examination, a positive apprehension sign is usually found.

Pathophysiology

An acute shoulder dislocation is often associated with damage to the glenoid labrum (the lip of cartilage which surrounds and deepens the glenoid; "Bankart lesion"), a compression fracture of the head of the humerus (the "Hill-Sachs lesion"), and/or damage to other soft tissues which usually aid in stabilizing the shoulder. This then predisposes to recurrent subluxation or repeated dislocations, which in turn can lead to degenerative change in the shoulder.

Treatment

A *strengthening program* supervised by a physical therapist may be helpful.
But referral to an orthopedist for consideration of any of a variety of *surgical procedures* is usually necessary.

SHOULDER TRAUMA

Always examine distal neurovascular status.
Fractures and dislocations of the shoulder are beyond the scope of this discussion. Rotator cuff tears are discussed above.

SHOULDER TRAUMA

Acromioclavicular Joint Sprain (Shoulder Separation, Shoulder Pointer) *(Common)*

Diagnosis

These are usually caused by a fall on the point of the shoulder.

Tenderness is found over the acromioclavicular joint (Figure 6-2).

X-rays should be done with a 5-to-10 lb weight dangling in each hand and comparison made to the uninjured side. The radiograph will be normal in a first-degree sprain, will show a slight degree of displacement (not more than the width of the end of the clavicle) in a second-degree sprain, and will show a significant upward displacement of the distal end of the clavicle in a third-degree sprain.

Pathophysiology

A first-degree sprain involves an incomplete tear of the acromioclavicular ligament without joint subluxation, and a second-degree injury more severe disruption of the tissues with some joint subluxation. In a third-degree sprain the coracoclavicular ligaments are torn as well.

Treatment

First-degree and mild second-degree sprains should be treated with a *sling* for one to three weeks until the pain is resolving, and then the exercises shown in Patient Handout 6-1 should be begun to restore strength and maintain range of motion.

Severe second-degree sprains and all third-degree sprains should be *referred* to an orthopedist for management.

Shoulder Strain *(Common)*

Diagnosis

Tenderness may occur in various locations depending on the specific tissues involved.

Pain on passive or resisted motion is also variable.

Pathophysiology

Musculotendonous injury.

Treatment

Ice and analgesics or antiinflammatories.

A sling may be used for a short time.

PATIENT HANDOUT 6-1

Shoulder Mobilization Exercises

These exercises are very important to prevent your shoulder from stiffening up, or to restore a full range of motion if some restriction has already occurred. Do them twice a day.

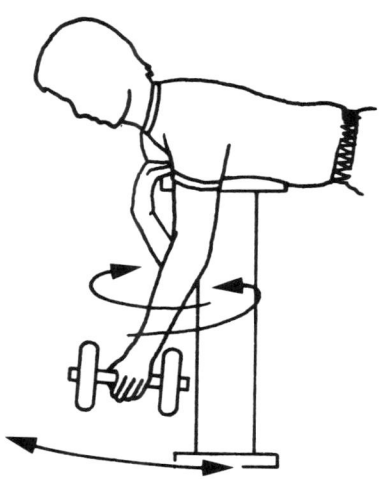 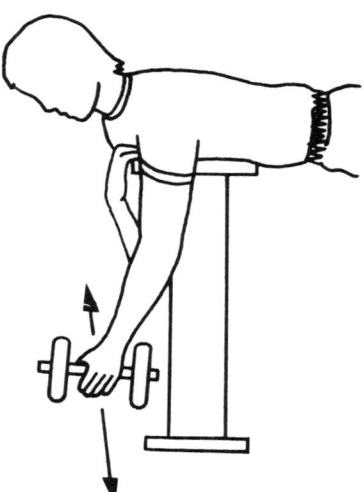

PENDULUM EXERCISE

Do this with a weight in your hand. Move your arm forward and back further and further with each swing until it's swinging as far as it will go; at that point, swing ten more times. Then do the same thing side-to-side. Then make circles in one direction, swinging wider and wider; then the same in the other direction.

PATIENT HANDOUT 73

WALL CLIMBING

Face the wall with your arm outstretched and your fingertips just touching; now without leaning your body, walk your fingertips up the wall as far as they will go. Repeat five times.

☐ Do the same thing with your side toward the wall and your arm elevated at your side. Do this one only if your doctor has checked the box.

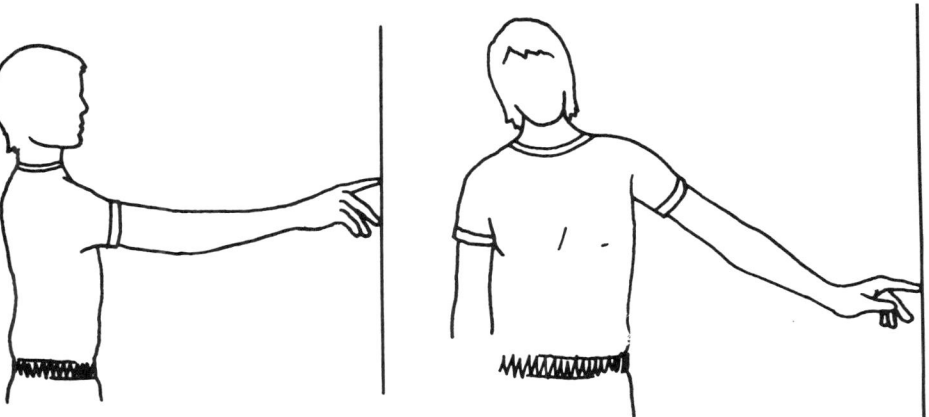

7

The Elbow

Elbow Pain	74
History	74
Examination	76
Lateral Epicondylitis	76
Medial Epicondylitis	79
Degenerative Arthritis of the Elbow	79
Elbow Swelling	80
Evaluation	80
Olecranon Bursitis	81
Elbow Trauma	81
Evaluation	81
Elbow Sprains	81

Elbow problems in *toddlers* and *throwers* are discussed in Chapter 19. Further details on therapeutic suggestions made in this chapter can be found in Chapter 3.

For the anatomy of the elbow see Figure 7-1.

ELBOW PAIN

History

Ask about the *duration* and *location* of the pain. (Occasionally a diffuse ache may be *referred pain* from cervical or intrathoracic disease; it is wise to keep this in mind).

What are *precipitating* and *relieving factors?*

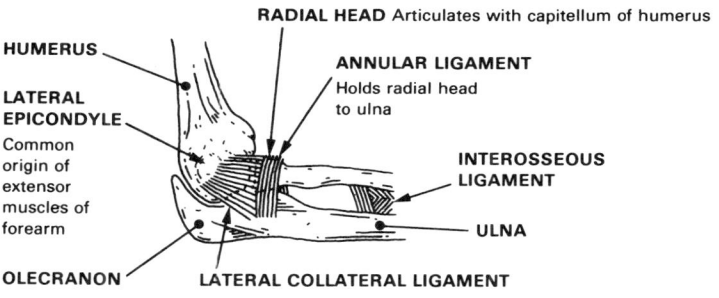
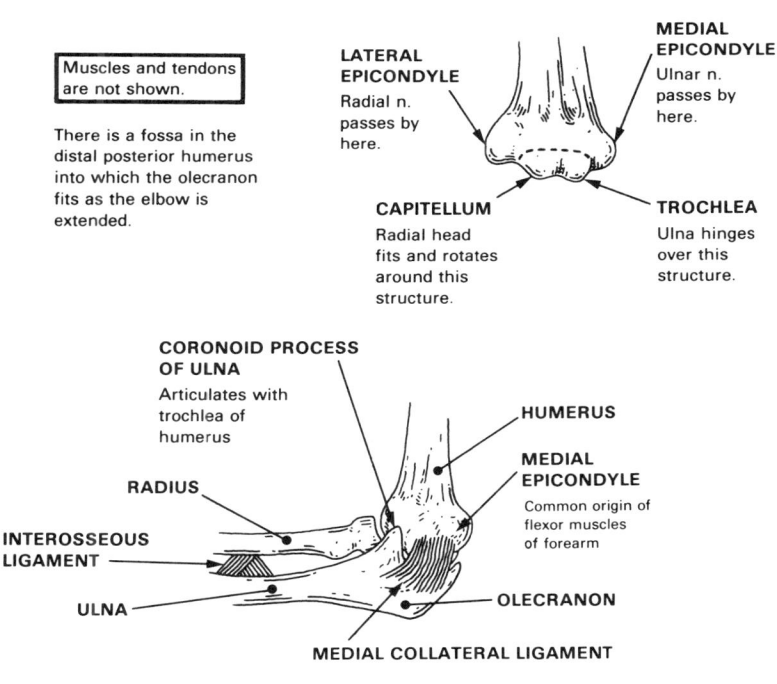

Fig. 7-1. Anatomy of the elbow.

Has there been any specific *trauma, strain,* or *overuse?*
Inquire about any *previous elbow problems,* diagnoses, and treatments.
Are any *other joints* involved, or is there a history of arthritis or gout?
Does the joint ever *lock?* (True locking indicates a loose body in the joint from previous injury, osteochondritis dissecans, or severe degenerative changes. Take an X-ray and, if symptoms are significant enough to warrant treatment, refer the patient to an orthopedist.)

Examination

Localize *tenderness* (Figure 7-2).
Look for *swelling* (Figure 7-3).
Of course, any *redness* or *warmth* implies an acute inflammatory process.
Test *range of motion*: pronation/supination (palm up/palm down; normal is 90° each way) as well as flexion/extension (normal is 0 to 160°). (*Note:* A flexion contracture, i.e., lack of full extension, is a nonspecific sign of long-term elbow disease and disuse.) Painful range of motion in the absence of acute trauma indicates an intra-articular process, most likely degenerative arthritis in chronic cases, crystals or infection in acute ones.
Check *distal pulses, strength,* and *sensation.*

Lateral Epicondylitis (Tennis Elbow) *(Very Common)*

Diagnosis

Tenderness is at or near the lateral epicondyle (Figure 7-2).
Confirm by pain referred there on resisted wrist extension and/or supination.
Elbow range of motion is essentially full and painfree. (Sometimes with severe epicondylitis, there may be some pain at extreme flexion or extension.)

Pathophysiology

Inflammation (and possibly some microtears) occurs where the common extensor tendon of the forearm attaches, precipitated by acute or chronic use of the extensor and supinator muscles of the forearm.
Common with various occupations as well as with racket sports.

Treatment

Have the patient *avoid* the responsible activity as much as possible.
A *wrist splint* is especially useful where the extensor muscle overuse is with wrist movement (as opposed to isometric): in a house painter, for example.
A *tennis elbow band* (a *non*elastic strip applied circumferentially around the proximal forearm) theoretically prevents excessive muscle contraction by constriction, and is more useful in isometric overuse of the muscle.

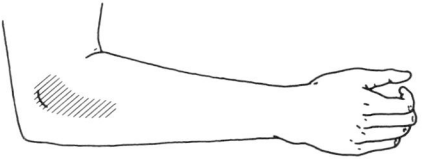

LATERAL EPICONDYLITIS
>   There will be pain here as you resist the patient's effort to extend (dorsiflex) his wrist and/or supinate his forearm.

MEDIAL EPICONDYLITIS
>   There will be pain here as you resist the patient's effort to flex his wrist and/or pronate his forearm.

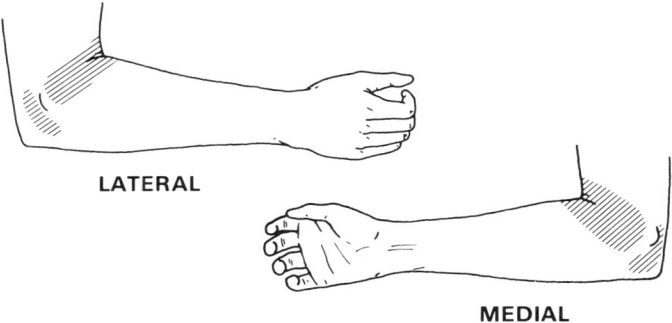

LATERAL

MEDIAL

An Intraarticular Process
>   Consider DEGENERATIVE ARTHRITIS in an older patient or one with previous trauma; if there is redness, swelling, or warmth, the joint must be tapped to rule out infection (See Fig. 7-3).

Fig. 7-2. Sites of tenderness: the elbow.

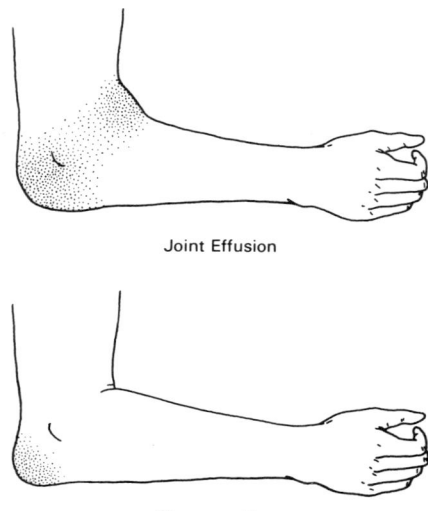

Joint Effusion

Olecranon Bursa

Fig. 7-3. Elbow swelling.

In those who play racket sports, advise a lighter racket with smaller grip and less string tension) using a two-handed back-hand, and not applying excessive backspin.

Note that immobilization of the elbow is irrelevant and useless.

Antiinflammatory medications may be helpful.

Steroid injection may be offered. Success rate is only fair; see Figure 7-4.

Advise the application of ice for twenty minutes at a time on a regular basis or prn. Some patients report better symptomatic relief with heat.

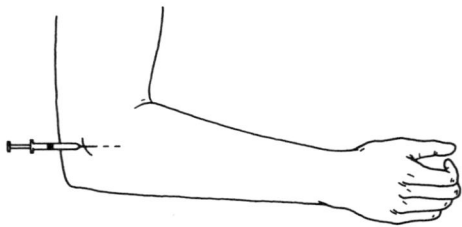

Fig. 7-4. Injection for epicondylitis. Prep the area and spray with ethyl chloride. Use a 23-gauge needle to inject a cc of steroid and a cc of local anesthetic in the manner shown. Injections may be repeated weekly or biweekly to a total of three or so. See pp. 22-24 for further details and precautions.

ELBOW PAIN

The program in Patient Handout 7-1 is useful for rehabilitation and to prevent recurrence once the acute episode has calmed. Some feel it should be started when pain is still present in long-term cases, to try to form a painfree scar at the muscle origin.

In patients who actually play tennis, recommend the program outlined in Patient Handout 7-2.

Surgery should be a last resort only, since most cases of epicondylitis improve in a few months. The usual procedure is a simple release of the inflamed tendon from the epicondyle, allowing healing in a lengthened position.

Note that some patients with what seems to be refractory epicondylitis may actually have the RADIAL TUNNEL SYNDROME, which is entrapment of the posterior interosseous nerve (a branch of the radial) as it passes under the supinator muscle. Tenderness will be localized over muscle, an inch or so distal, and a bit toward the antecubital area, from the epicondyle. Pain will be referred to that point with resisted middle finger extension and forearm supination (done with the elbow straight). Surgical release has a very good success rate.

Medial Epicondylitis *(Common)*

Diagnosis

    Tenderness is at the medial epicondyle (Figure 7-2).
    Confirm by pain referred there on resisted wrist flexion and/or pronation.
    There is usually full and pain-free range of motion of the elbow.

Pathophysiology

    Analogous to lateral epicondylitis, but with overuse of the flexor and pronator muscles of the forearm.
    Common with throwing and golfing.
    A short discussion of elbow problems related to throwing is in Chapter 19.

Treatment

    Analogous to lateral epicondylitis.

Degenerative Arthritis (Osteoarthritis) of the Elbow
(Uncommon)

Diagnosis

    Chronic pain is increased after use.
    Found in older patients or those with previous significant trauma.
    There is variable diffuse tenderness and variable pain with range of motion.

There may be slight swelling in an acute exacerbation, but if it is red or warm it must be tapped to rule out an infection.

The X-ray may be normal early (when changes are limited to joint cartilage), but it is characteristic later. (But remember that the presence of degenerative changes on X-ray does not prove that any particular patient's symptoms are due to those changes.)

Pathophysiology

See Chapter 22.

Treatment

See Chapter 22.
Technique of injection is shown in Figure 7-5.

ELBOW SWELLING

Evaluation

Swelling of the olecranon bursa must be differentiated from swelling of the joint itself, which implies an intraarticular process. See Figure 7-3.

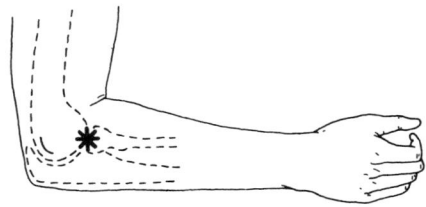

Fig. 7-5. How to aspirate or inject an elbow joint: (a) Feel the tip of the lateral epicondyle with the elbow at 90° of flexion. Now move just distal, to it (toward the hand) and feel the joint line between the humerus and the radial head. (b) Prepare the site and use sterile technique throughout. After tehyl chloride spray, enter the joint using a 22- or 20-gauge needle and aspirate. (c) If steroid is to be injected, use a hemostat to hold the needle bevel and switch syringes to one containing 2 to 4 cc of steroid. (Never inject steroid if the synovial fluid is cloudy or if there is other reason to suspect infection.) Steroid can be injected the same way, without aspiration, using a 22-gauge needle. Be sure to see pp. 12 and 22-24 for discussion of synovial fluid analysis and of precautions and risks *re* steroid instillation.

Olecranon Bursitis *(Common)*

Diagnosis

A painless or slightly painful swelling is localized over the olecranon (Figure 7-3).

If there is redness, warmth, significant tenderness, or a nearby break in the skin, it must be tapped to rule out infection.

Pathophysiology

Fluid accumulates in the inflamed bursa, usually secondary to leaning on the elbow (especially on a hard surface).

Occasionally infection, and rarely gout, may be responsible.

Treatment

The patient must *avoid* leaning on the elbow; a *pad* may be used to prevent recurrence if leaning is unavoidable.

Some feel that immobilizing the elbow in a *sling* for a short time helps by avoiding rubbing the bursa with elbow flexion and extension.

Heat may help.

Drainage may be offered; the fluid usually reaccumulates very quickly unless some *steroid* is instilled after drainage, but even this often does not work. Technique is simply to enter the center of the swelling with a #18 needle (with a large syringe) after sterile preparation and ethyl chloride spray; never inject steroid if the fluid drained appears cloudy or purulent. Remember, the bursa must be aspirated if the signs of infection described above are present. Apply a *compressive bandage* after drainage.

Surgery may be done to remove a recurrently or chronically inflamed bursa (which may have a lumpy feeling to it, by the way).

ELBOW TRAUMA

Evaluation

Always examine distal neurovascular status.

Note: If there is significant pain or swelling in the elbow after direct or indirect trauma (such as a fall on the outstretched hand), or a positive fat pad sign (Figure 7-6) on X-ray, suspect a fracture even if not visualized, and refer the patient to an orthopedist.

Elbow Sprains *(Uncommon)*

Diagnosis

Usually caused by a hyperextension injury.

There is diffuse soft tissue swelling and tenderness.

No fracture or fat pad sign (Figure 7-6) is seen on X-ray.

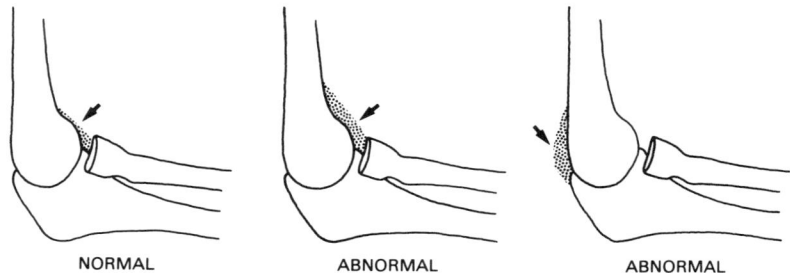

Fig. 7-6. The fat pad sign on elbow X-ray. A small anterior fat pad is normal. But a large one, or any visible posterior fat pad, is indicative of an intraarticular fracture even if the fracture is not visualized. (And they can be notoriously hard to see.)

Treatment

If severe, referral to an orthopedist is best.

If mild with a negative X-ray, treat with *ice* intermittently for the first 48-72 hours, with *analgesia* and *rest* (including use of a sling).

Early active range of motion should be instituted to prevent a flexion contracture. When to begin depends on clinical progress.

PATIENT HANDOUT 7-1

Rehabilitation Exercises for Epicondylitis

Put your arm on a table with your wrist hanging over the edge, holding something in your hand that weighs a couple of pounds. Now *slowly* straighten your wrist and bend it upwards; then slowly turn it palm up. Now *slowly* turn it back palm down and lower it. Do 15 repetitions twice a day. Increase the weight by a couple of pounds every few days, until you are up to 10 or 15 pounds.

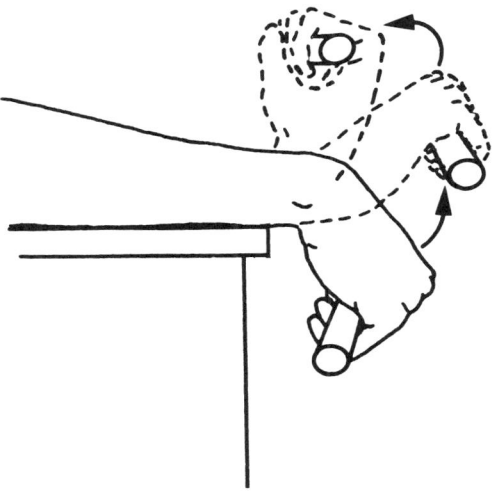

PATIENT HANDOUT 7-2

Tennis Elbow

This common affliction occurs because of the accumulation of stress and minor damage where the tendon from the forearm muscles attaches at the outside of the elbow. The area then becomes inflamed and painful. There are many things you should do to help alleviate the problem:

1. Take two aspirin one half hour before playing. This should reduce the amount of inflammation that will occur. Of course you can also take aspirin for the pain as needed (unless your doctor has given you a different medicine to take).
2. Make sure you warm up slowly and gradually before getting competitive and hitting the ball hard.
3. An elbow sleeve made out of covered neoprene is widely available and can help keep the area warm and flexible during play.
4. Be sure you choose a racket that will reduce the stress on your forearm muscles. It should have:
   - a proper size grip (many players use one that is too large)
   - a large "sweet spot" to cut down the chance of mishitting (which is when your arm is extra-stressed)
   - a lighter weight
   - proper string tension . . . not too tight (50-55 pounds is usually best)
   - metal rackets transmit a lot of vibration and should be avoided.
5. Hitting the backhand incorrectly is a principal cause of tennis elbow. Players who hit with a "leading elbow" tend to have a lot of trouble. Using a two-handed backhand or getting a good coach can help a lot.
6. Some players find that wearing a non-elastic "tennis elbow band" below the elbow is helpful.
7. When you finish playing, be absolutely sure to put an ice pack on the sore area for twenty minutes or so. You can also apply ice at other times that the elbow is bothering you.
8. A series of exercises to strengthen your forearm muscles is very important for prevention. Your doctor may give you these on a separate sheet.

PATIENT HANDOUT

9. Despite all of this, if your elbow is very sore it will be necessary to rest it for a while to allow some healing before you go back to play, and then to build up your playing time slowly enough to prevent the problem from getting worse again . . . in other words, listen to your body!

Good luck!

8

The Wrist

Wrist Pain 86
    History 86
    Examination 87
    Dequervain's Tendonitis 88
    Wrist Tendonitis 88
    Degenerative Arthritis of the Carpometacarpal Joint of the Thumb 90
    Degenerative Arthritis of the Wrist 91
Wrist Lumps 92
    Ganglia 92
Wrist Trauma 94
    Evaluation 94
    Wrist Sprain 94

For the anatomy of the wrist see Figure 8-1.

WRIST PAIN

History

Ask about the *duration* and *location* of pain.
What are *precipitating* and *relieving factors*?
Has there been any specific *trauma, strain,* or *overuse*?
Inquire about any *previous wrist problems,* diagnoses, or treatments.
Are any *other joints* involved? Is there a history of arthritis or gout?

WRIST PAIN

1. **NAVICULAR (SCAPHOID)**
   Most often fractured
2. **LUNATE**
   Second most commonly fractured
3. **TRIQUETRUM**
4. **PISIFORM**
   Volar to other bones
5. **TRAPEZIUM**
   (GREATER MULTANGULAR)
6. **TRAPEZOID**
   (LESSER MULTANGULAR)
7. **CAPITATE**
8. **HAMATE**

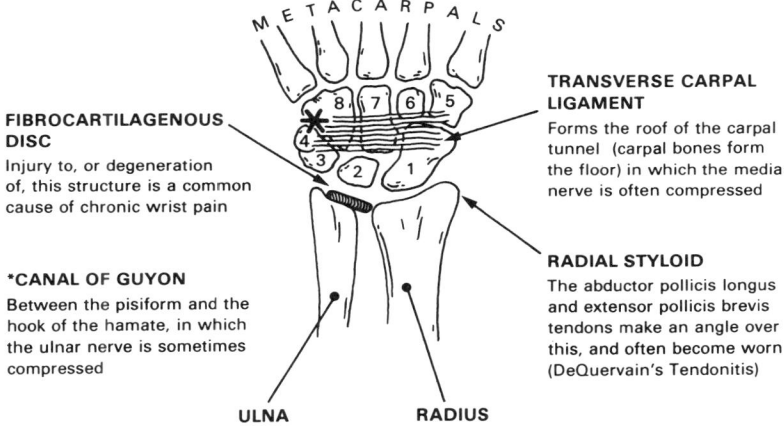

FIBROCARTILAGENOUS DISC
Injury to, or degeneration of, this structure is a common cause of chronic wrist pain

*CANAL OF GUYON
Between the pisiform and the hook of the hamate, in which the ulnar nerve is sometimes compressed

TRANSVERSE CARPAL LIGAMENT
Forms the roof of the carpal tunnel (carpal bones form the floor) in which the median nerve is often compressed

RADIAL STYLOID
The abductor pollicis longus and extensor pollicis brevis tendons make an angle over this, and often become worn (DeQuervain's Tendonitis)

Many tendons pass over the wrist. Those on the dorsal surface are covered by a synovial sheath.
There are also a multitude of ligaments (not shown) interconnecting the bones.

Fig. 8-1. Anatomy of the wrist.

If pain radiates into the hand or up the arm or is associated with numbness or paresthesias, see Chapter 5.

Examination

Localize *tenderness* (See Figure 8-2).
Any *generalized* swelling in the absence of trauma implies an intraarticular process, which if acute *must be tapped* to rule out infection.
Any *redness* or *warmth* of course also implies an acute inflammatory process.

88   8. THE WRIST

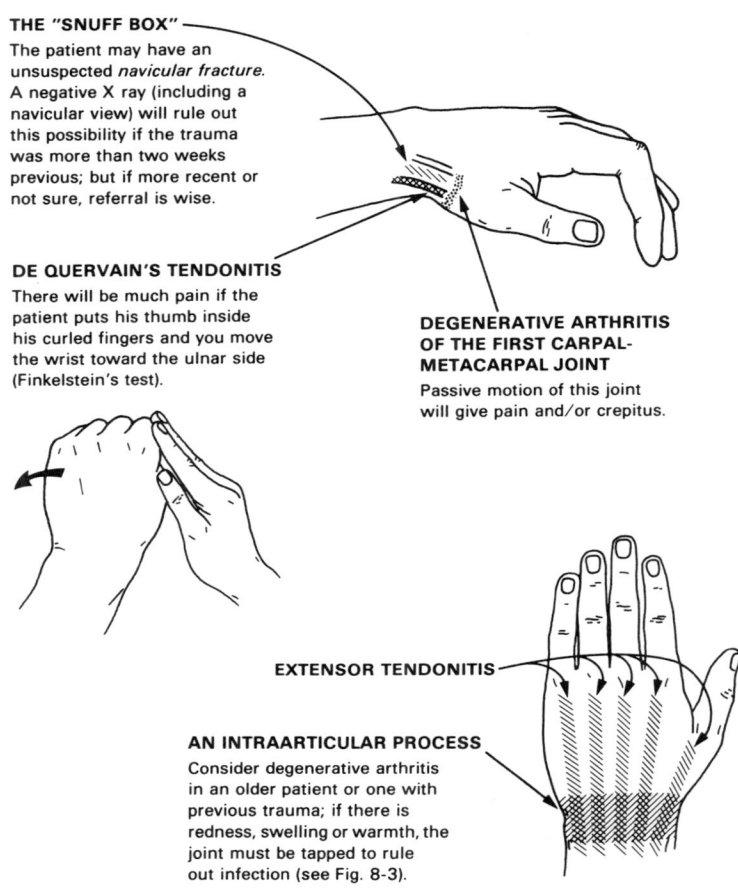

Fig. 8-2. Sites of tenderness: the wrist.

Test *range of motion* (normal is 80° flexion, i.e., turned toward the palm, 70° extension, 30° ulnar deviation, and 20° radial deviaton).

Check *strength, sensation* and *distal capillary filling*.

De Quervain's Tendonitis *(Very Common)*

Diagnosis

Tenderness is found over the radial styloid and/or the tendons that pass over it (Figure 8-2).

WRIST PAIN

Confirm by eliciting much pain with Finkelstein's test (Figure 8-2), and often also with pain on resisting the patient's attempt to extend his thumb.

Pathophysiology

Inflammation of the abductor pollicis longus and extensor pollicis brevis tendons (which move in the same sheath) results from friction where they form a sharp angle over the radial styloid.

Common with knitting, the use of a wrench, holding a baby, and various other activities.

Treatment

Advise the patient to *avoid* the responsible activity as much as possible.

Use a *wrist splint* (Figure 8-3). (This is somewhat helpful but incompletely so, since the thumb is not immobilized. Also, be sure part of the splint is not exerting pressure over the sore area.)

Instead, a short arm *cast* with a thumb extension may be resorted to in order to achieve adequate immobilization.

Antiinflammatory medications and *heat* are helpful.

Steroid injection may be offered. Success rate is quite good. See Figure 8-4.

In prolonged or recurrent cases when conservative measures and injection have failed, referral to an orthopedist for *surgery* (in which the tendon sheath is split longitudinally) should be considered.

Wrist Tendonitis *(Common)*

Diagnosis

Linear tenderness is found over a specific tendon.

Range of motion is painful only when stressing the involved tendon.

There is *no* redness or warmth. (Infectious tenosynovitis is not uncommon here; if suspected, immediate referral to an orthopedist is mandatory.)

Pathophysiology

Tendon inflammation occurs from minor strain or overuse.

Treatment

Use a wrist splint (Figure 8-3).

Antiinflammatory medications and *heat* are helpful.

Steroid injection around (not into) the involved tendon may be offered in long-term cases. See pp. 24–26 for details and precautions.

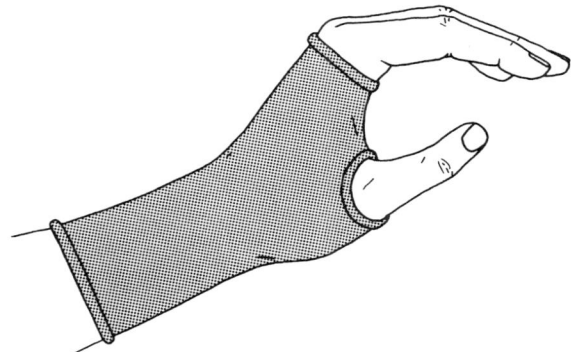

Fig. 8-3. Wrist splint.

Degenerative Arthritis of the Carpometacarpal Joint of the Thumb *(Common)*

Diagnosis

Tenderness is found around that joint (Figure 8-2).
There is pain on motion of that joint, often with crepitus.

Pathophysiology

Accelerated wear and tear occurs here from carrying objects in a pinch between thumb and fingers.
Common in mail carriers.

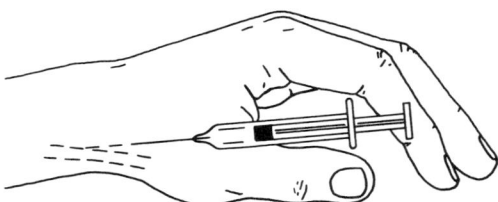

Fig. 8-4. Injection For DeQuervain's tendonitis. Prep the area and spray with ethyl chloride. Enter the tendon sheath with a 25-gauge needle as shown. (One way to do this is to enter the tendon and slowly pull back until the needle pops out.) Inject half a cc each of steroid and of local anesthetic without epinephrine. If you are in the sheath you will see the fluid distend it in a linear pattern. Never inject into the tendon itself. See pp. 24-26 for further details and precautions.

WRIST PAIN

Treatment

The patient should *avoid* the responsible activity as much as possible.

Steroid injection is usually helpful. See Figure 8-5.

If the patient declines injection, oral *antiinflammatory medication* may be used.

A *splint* to immobilize the joint can be fabricated for the patient by an occupational therapist and can be quite helpful.

Degenerative Arthritis of the Wrist (Diffuse) *(Uncommon)*

Diagnosis

Chronic pain is worse after use.

Found in older patients or those with previous trauma.

There is variable diffuse tenderness and variable pain with range of motion.

There may be slight swelling during an acute exacerbation, but if it is red or warm it must be tapped to rule out infection.

Pathophysiology

See Chapter 22. The process may involve only certain articulations within the wrist.

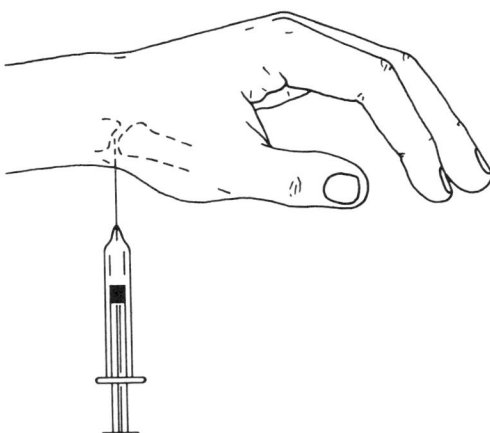

Fig. 8-5. Injecting the CMC joint of the thumb. (a) Find the joint line with your thumb by putting your fingers on the thenar eminence and moving the metacarpal. (b) Prep. the area and spray with ethyl chloride. (c) Inject into the dorsal radial aspect with a 25-gauge needle ½ cc each of steroid and local anesthetic without epinephrine. See p. 24 for further details and precautions.

8. THE WRIST

Treatment

See Chapter 22.
Technique of injection is shown in Figure 8-6.

WRIST LUMPS

Ganglia *(Very Common)*

Diagnosis

The lumps can occur at any location around the wrist, and may come and go. Increase in size is often related to joint usage.

They are sometimes painful, and rarely can cause symptoms of nerve compression.

On examination, a smooth, rounded cystic lump is found, which is nonpulsatile and usually ½ to 2 cm in size.

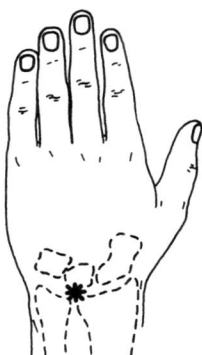

Fig. 8-6. How to aspirate or inject a wrist joint. (a) Feel the radius and ulna on the dorsal aspect of the wrist; there is a palpable depression between their distal ends. (b) Prepare the site and spray with ethyl chloride. Use sterile technique throughout. Enter the joint perpendicularly using a 22-gauge or 20-gauge needle, and aspirate. (c) If steroid is to be injected, use a hemostat to hold the needle bevel and switch syringes to one containing 2 cc of steroid. (Never inject steroid if the synovial fluid is cloudy or if there is other reason to suspect infection.) Steroid can be injected the same way, without aspiration, using a 22-gauge needle. Be sure to see pp. 14 and 22 for discussion of synovial fluid analysis and of precautions and risks *re* steroid instillation.

WRIST LUMPS

If the lump is *progressively growing* or *not obviously cystic,* the patient should be *referred* for *biopsy* to rule out a possibly neoplastic lesion, though this would be extremely rare.

Pathophysiology

The cyst arises from a tendon sheath or from the joint synovium.

Treatment

If it is not progressively growing by history and is obviously cystic on examination, and if it does not bother the patient, no treatment is necessary.

A wrist splint can be symptomatically helpful if the patient is not interested in more definitive treatment.

The ganglion may be *aspirated*; see Figure 8-7. Recurrence is common.

If this method is unsuccessful or if the ganglion recurs, the patient may be offered referral to an orthopedist for *surgical removal*. Excision can be somewhat complex, especially in those ganglia that arise from the joint synovium, due to the necessity for tracing and removing multiple long roots of the ganglion. Failure to do this completely accounts for some cases of postoperative recurrence.

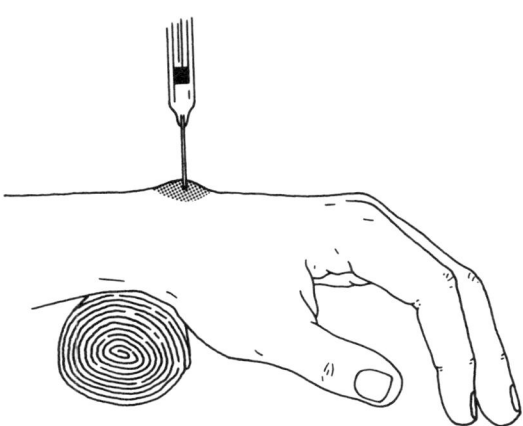

Fig. 8-7. Aspirating a ganglion. (a) Flex the wrist over a pillow or roll as shown. Make sure the mass is not pulsatile. (b) After prep and ethyl chloride spray, enter the ganglion with an 18-gauge needle and aspirate. (c) Whether or not the thick gelatinous contents of the ganglion are obtained, use finger pressure to flatten the cyst into disappearance. Multiple punctures and/or the installation of a small amount of steroid may decrease the recurrence rate. If the cyst cannot be flattened, it may not be a ganglion, and must be referred for excisional biopsy.

8. THE WRIST

WRIST TRAUMA

Evaluation

Those who make the often-heard statement "there is no such thing as an acute wrist sprain" may be carrying things a bit far, but carpal bone *fractures,* especially of the navicular (scaphoid) bone, are notorious for *not showing up on early X-rays* and for leading to serious consequences if not adequately treated. The navicular in particular is prone to avascular necrosis after a fracture, due to its tenuous blood supply.

Furthermore, certain specific ligamentous injuries of the wrist need early specific treatment if disability is to be minimized.

Therefore any acute wrist injury with any swelling or significant tenderness [or even mild tenderness if located over the navicular (Figure 8-2)] should be referred to an orthopedist.

Always check distal neurovascular status.

Fractures and dislocations are beyond the scope of this book.

Wrist Sprain *(Common)*

Diagnosis

See the caution above.
Usually due to a hyperextension injury.

Treatment

Use a *wrist splint, analgesics,* and for the first 48 hours, *cold* and *elevation.*

9

The Hand

Diffuse Nonlocalized Hand Pain 98
    Evaluation 98
Atraumatic Pain in the Joints of the Hand and Fingers 98
    Evaluation 98
Snapping Finger 99
    Trigger Finger 99
Fractures 99
    Types 99
    Tuft Fracture 100
Dislocations 101
    PIP Joint 101
Sprains of the MCP Joint of a Finger 102
Sprains of the MCP Joint of the Thumb 102
    Gamekeeper's Thumb 103
Sprains of the PIP Joint 103
    Evaluation 103
    Volar Plate Injury 103
    Central Extensor Slip Tear 104
    Collateral Ligament Sprain 105
Sprains of the DIP Joint 105
    Evaluation 105
    Mallet Finger 106
Contractures 107
Lumps 107
Infections 108

Further details on therapeutic suggestions made in this chapter can be found in Chapter 3.

    For the anatomy of the hand see Figure 9-1.

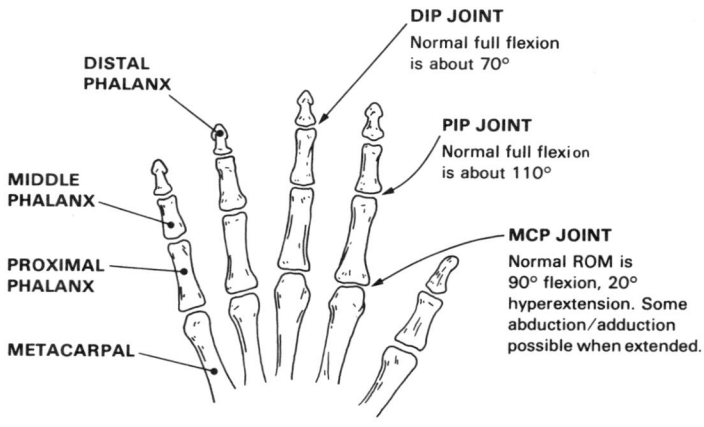
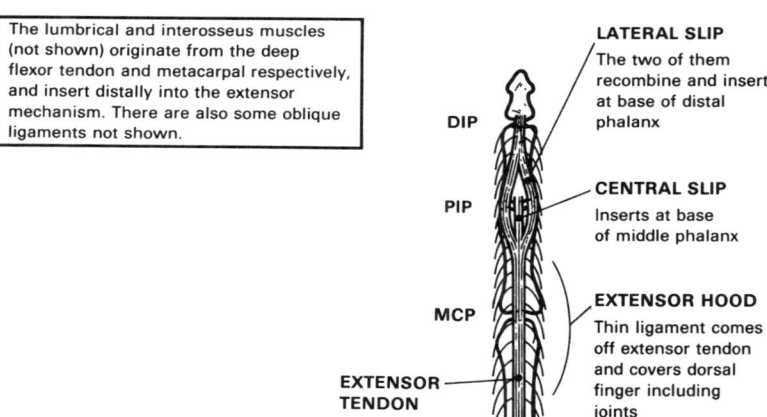

Fig. 9-1. Anatomy of the hand.

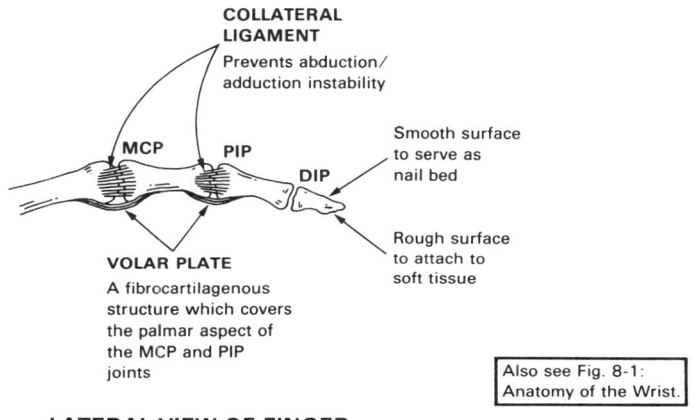

LATERAL VIEW OF FINGER

Note: The anatomy of the hand and digits is extremely complex; many fascial layers, septae, ligaments, tendons, and intrinsic muscles are not shown here. Nor, obviously, are nerves and vascular structures.

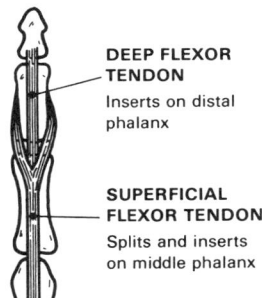

PALMAR VIEW OF FINGER SHOWING FLEXOR TENDONS

Fig. 9-1. Continued.

DIFFUSE NONLOCALIZED HAND PAIN

Evaluation

A neurovascular examination is essential.

Diffuse hand aching, without localized tenderness on examination, especially if worst at night, is suggestive of CARPAL TUNNEL SYNDROME, which is discussed in Chapter 5. Other radiated pain (from the neck or thoracic outlet or even cardiopulmonary disease) should also be considered. See Chapter 5.

Diffuse hand or finger pain which comes on with exposure to cold and is relieved with rewarming, especially if associated with color change, suggests *Raynaud's Phenomenon*. The reader is referred to a textbook of rheumatology or internal medicine for further discussion of this condition.

ATRAUMATIC PAIN IN THE JOINTS OF THE HAND AND FINGERS

Evaluation

Acute atraumatic monoarthritis in the hand has the same significance as elsewhere, with the leading possibilities being infection or crystalline disease; the joint must be tapped, especially to rule out the former possibility.

Three types of arthritis commonly affect the joints of the hand and fingers: OSTEOARTHRITIS, RHEUMATOID ARTHRITIS, and PSORIATIC ARTHRITIS.

OSTEOARTHRITIS more commonly affects the DIP and PIP joints; morning stiffness lasts only a few minutes; examination will commonly reveal hypertrophy at the DIP joints (Heberden's nodes) and less commonly, the PIP joints (Bouchard's nodes). Redness and warmth are uncommon; X-ray will show joint space narrowing and hypertrophic changes of bone; blood tests such as the ESR and rheumatoid factor will be normal. See Chapter 22.

One clinical type of PSORIATIC ARTHRITIS involves primarily the DIP joints. Involvement is usually asymmetric; the joint disease may become manifest before any skin lesions are apparent, but usually involved fingers will be found on close inspection to have pitting of their nails also. X-ray will show some bony erosion, especially in the tufts of the distal phalanges; the sedimentation rate may be elevated, while the rheumatoid factor will be negative. See Chapter 23.

RHEUMATOID ARTHRITIS in the hands involves mainly the PIP and MCP joints. Morning stiffness will be a significant complaint; systemic symptoms can be present. Inflammation will usually be evident in acute cases, and synovial thickening will be palpable in chronic ones. Various deformities can be found in more advanced cases. The X-ray may be normal early, but will eventually begin to show erosions of bone at the joint margins. The ESR will be elevated, and the rheumatoid factor will usually be positive. See Chapter 23.

SNAPPING FINGER

Trigger Finger (or Trigger Thumb) *(Common)*

Diagnosis

The patient will complain, and examination will confirm, that when the involved digit is flexed there is resistance to reextension in mid-arc which sometimes can only be overcome by extending the digit passively with the other hand.

Reextension is accompanied by a palpable and sometimes audible pop.

There will often be tenderness (and sometimes a palpable lump as well) at the base of the flexor tendon sheath.

Pathophysiology

A thickening develops in the flexor tendon of the digit, and resistance to entrance into the base of the flexor tendon sheath is encountered as the digit is extended.

The thickening is most usually a consequence of overuse but can be related to rheumatic disease.

Treatment

Success rate of *steroid injection* is good if done properly. See Figure 9-2. It may have to be repeated at intervals.

Referral for *surgery* (in which the tendon sheath is split) is the only other treatment option available.

FRACTURES

Types

Fractures of the tuft of the distal phalanx are discussed below.

Avulsion fractures at the base of the phalanges are treated as the corresponding sprain (Table 9-1), unless the fragment involves more than about one-fourth of the joint surface, in which case the patient should be referred to an orthopedist.

The following fractures require reduction or at least the application of plaster; the reader is advised to refer to a textbook of orthopedics or to refer the patient for proper treatment.

1. Fractures of the metacarpals.

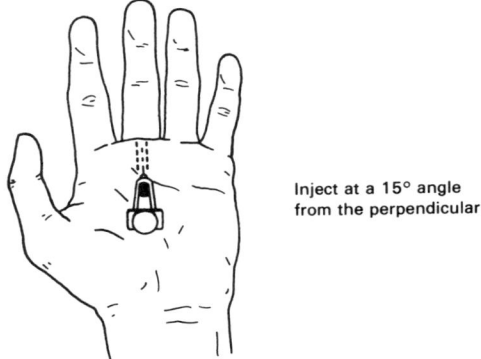

Inject at a 15° angle from the perpendicular.

Fig. 9-2. Injection for trigger finger. Find the tender area at the base of the flexor tendon sheath (just distal to the distal palmar crease for the fingers, at the level of the MCP crease for the thumb). After prep and ethyl chloride spray, enter the tendon as shown with a 25-gauge needle and slowly withdraw until you just feel it come out of the tendon. Now inject about a quarter of a cc each of steroid and local anesthetic without epinephrine (mixed in the same syringe, of course). Never inject against resistance. See Chapter 3 for further details and precautions.

2. Transverse, oblique, or spiral fractures of the proximal or middle phalanges.
3. Transverse fractures of the proximal portion of the distal phalanx.

Tuft Fracture of the Distal Phalanx *(Common)*

Diagnosis

These are usually sustained by crushing-type injuries.

If there is a transverse fracture of the proximal end of the distal phalanx, refer the patient for treatment.

Table 9-1   Avulsion Fractures

Avulsion	Corresponding sprain
Corner of base of middle phalanx	Collateral ligament
Volar base of middle phalanx	Volar plate injury
Dorsal base of middle phalanx	Central extensor slip tear
Volar base of distal phalanx	Flexor profundus tear (REFER)
Dorsal base of distal phalanx	Mallet finger

DISLOCATIONS

Management

Tuft fractures, even when severely comminuted, can be treated simply by preventing further trauma with a *splint* as shown in Figure 9-3.

DISLOCATIONS

Dislocations of the MCP joint and their reduction can lead to complications; their management is best left to the orthopedist.

PIP Joint *(Common)*

Diagnosis

This is usually due to a hyperextension injury.
An X-ray should be done to rule out an associated fracture.

Management

Reduction can be done as shown in Figure 9-4. Anesthesia is usually unnecessary.

After reduction, check to be sure that there is not hyperextensibility or volar

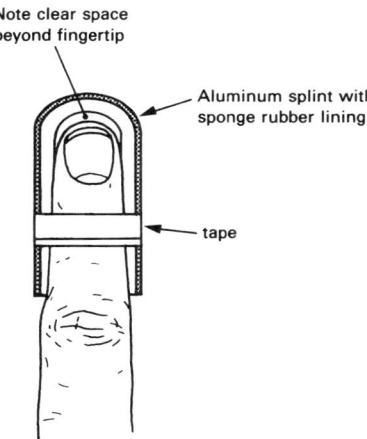

Fig. 9-3.   Splint for a tuft fracture.

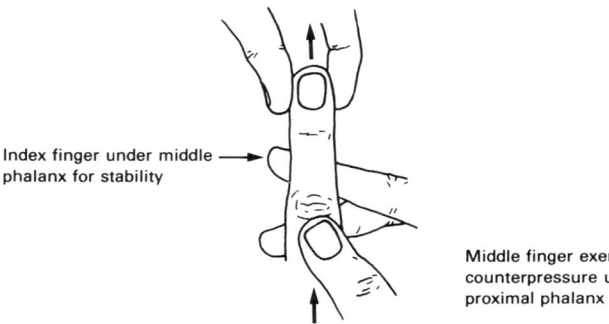

Fig. 9-4. Reduction of dislocation of the PIP joint.

midline tenderness, or conversely, lack of full active extension of the joint. If any of these are found, the joint should be treated as for a VOLAR PLATE INJURY or CENTRAL EXTENSOR SLIP TEAR, respectively.

If the findings in the previous paragraph are not present, *splint* the joint at about 40° of flexion (Figure 9-6) for about ten days.

Ice packs will be helpful for the first several days.

SPRAINS OF THE MCP JOINT OF A FINGER

Sprains of these joints often involve at least some tearing of the extensor hood. This will manifest itself by the patient's inability to extend the joint actively from a fully flexed position, often with visible or palpable slippage of the extensor tendon off the dorsum of the joint. All significant injuries to this joint are best managed by the orthopedist.

SPRAINS OF THE MCP JOINT OF THE THUMB

These are most significant when they involve the ulnar collateral ligament of the joint.

SPRAINS OF THE PIP JOINT

Sprain of the Ulnar Collateral Ligament of the MCP Joint of the Thumb (Gamekeeper's Thumb) *(Common)*

Diagnosis

The patient usually gives a history of falling onto the inside of the thumb with the thumb held away from the fingers.

Especially common in skiers and football players.

Management

These injuries can lead to significant disability if inadequately treated, since stability of pinch can be lost. Therefore, if there is any significant tenderness or any suggestion of instability (see p. 104), or an avulsion fracture on X-ray, the patient should be *referred* to an orthopedist for management (which consists of several weeks in a thumb spica cast for stable injuries, or surgical repair for unstable ones).

SPRAINS OF THE PIP JOINT

Evaluation

If there is tenderness over the volar base of the middle phalanx, or if the joint can be passively *hyperextended,* or if there is an avulsion fracture at the volar base, the patient has a VOLAR PLATE INJURY.

If the patient cannot *fully* actively extend the PIP joint, he must be considered to have a CENTRAL EXTENSOR SLIP TEAR. The classic "boutonniere" deformity does not develop until later.

If the above findings are absent, consider the patient to have a COLLATERAL LIGAMENT SPRAIN.

Volar Plate Injury *(Uncommon)*

Diagnosis

The history will usually be of a hyperextension injury.

There will be tenderness over the volar PIP joint.

The PIP joint can be passively hyperextended (when compared to the other hand).

Management

If there is an avulsion chip which is at all large (more than about 15% of the joint) or displaced (the so-called WILSON'S FRACTURE), the patient should be referred to the orthopedist. Internal fixation may be necessary.

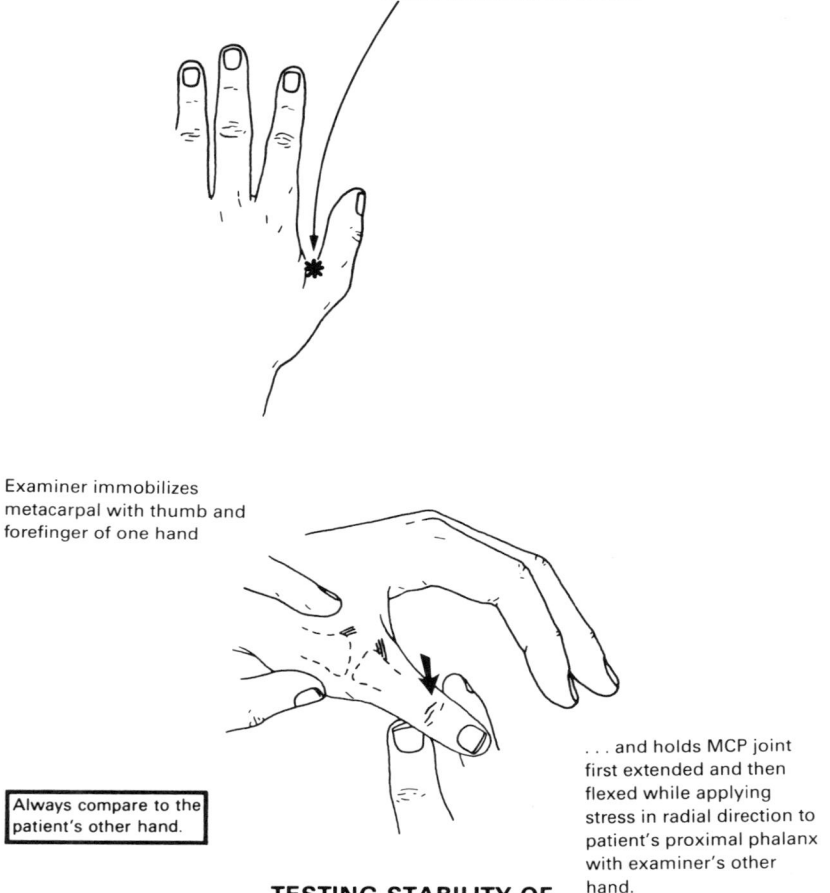

Fig. 9-5. Gamekeeper's thumb.

Otherwise the injury should be splinted in about 30° flexion for ten days, and then left in a 30° extension block splint for an additional ten days (Figure 9-6). During the latter period active motion in the splint is encouraged. Passive and active range of motion must be faithfully performed for several weeks after the splint is removed to prevent subsequent stiffness.

Central Extensor Slip Tear *(Uncommon)*

Diagnosis

Usually a history of sudden resisted flexion will be given.
The patient is unable actively to extend completely the joint.

SPRAINS OF THE DIP JOINT

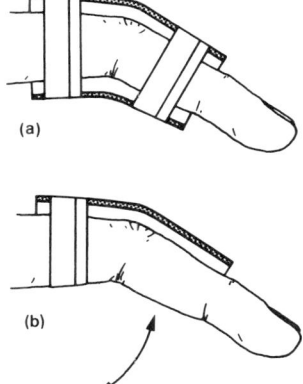

Fig. 9-6. (a) 30° to 40° flexion splint for the PIP joint. (b) 30° Extension block splint.

Management

The joint should be splinted in hyperextension (see Figure 9-7) for three weeks, and then active motion begun. Inadequate or delayed treatment will lead to the development of the classic "boutonniere" deformity, which must then be surgically corrected. Consider early orthopedic consultation.

Collateral Ligament Sprain *(Very Common)*

Diagnosis

Usually the mechanism is axial trauma (the "jammed finger").
The findings of VOLAR PLATE INJURY or CENTRAL EXTENSOR SLIP TEAR are absent.
Tenderness is maximal on the side of the joint.
Check for lateral stability with the joint in 15° of flexion.

Management

If there is an avulsion fracture of more than about 15% of the joint surface or gross instability, *refer* the patient to an orthopedist.
Otherwise, *splint* the joint in about 40° *of flexion* for 10 days, and then begin active motion.
Ice packs are helpful acutely.
Pain is often prolonged, and some soft tissue thickening may be permanent.

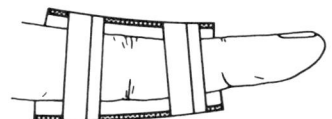

Fig. 9-7. Extension splint for the PIP joint.

SPRAINS OF THE DIP JOINT

Evaluation

See if the patient is able to flex the joint actively. If not, he probably has an avulsion of the flexor profundus tendon and should be sent to the orthopedist for repair.

Check to see if there is a loss of full active extension at the joint. If so, see MALLET FINGER, below.

If X-ray is normal with neither of the above findings, no treatment is necessary.

Mallet Finger (Baseball Finger, Dropped Finger) *(Common)*

Diagnosis

Usually the history is one of a blow on the dorsal part of the distal phalanx forcing sudden flexion.

The patient will be unable to extend the DIP joint fully.

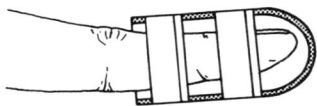

The DIP joint must be held in full extension or slight hyperextension.

Fig. 9-8. Mallet finger splint. If in hyperextension, be careful to avoid blanching of the skin over the dorsal joint, which could lead to sloughing. Various commercial splints are available for the treatment of this injury.

Management

If there is a large (greater than about one-fourth of the joint surface) avulsion fragment, *refer* the patient to the orthopedist.

Otherwise, *splint* the joint in *hyperextension* for six weeks. The trick is to impress upon the patient the need never to allow the fingertip to fall into flexion even for a moment until treatment is complete. See Figure 9-8. If treatment is delayed, the chance for successful nonsurgical healing is decreased.

CONTRACTURES

Obviously contractures of the various joints can occur with previous trauma or with denervation. DUPUYTREN'S CONTRACTURE is a thickening of the palmar fascia that eventually leads to flexion contracture of the fourth and/or fifth

HAND LUMPS

digits. It can be associated with cirrhosis of the liver, but is usually idiopathic. Treatment is by referral for surgical release when it advances to the point that the patient can't fully extend the involved digit.

HAND LUMPS

Evaluation

A tiny hard lump in the midline of the volar base of a digit is usually a small *sesamoid* bone in the flexor tendon sheath.

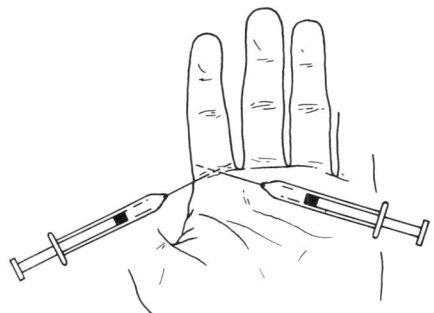

Fig. 9-9. Anesthezing a finger. After prep and ethyl chloride spray, inject about a cc of local anesthetic *without* epinephrine toward the medial and lateral sides of the finger from a point in the midline of the proximal crease. Use a 25-gauge needle. It usually takes about ten minutes for the anesthetic to take effect.

An extremely tender nodule following a laceration is probably a *neuroma*, which must be treated surgically.
GANGLIA can occur on the hand; see discussion of this topic in Chapter 8.
Other lumps (which are not benign-appearing exostoses on X-ray) should be referred for biopsy.

HAND INFECTIONS

Methods for anesthetizing a finger and draining a PARONYCHIA (subcuticular infection) are described in Figures 9-9 and 9-10. Infections of any of the various other spaces of the hand are potentially disastrous and should be referred STAT to a hand surgeon. ANY INFECTION following a HUMAN BITE/TOOTH WOUND should also be referred.

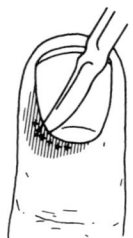

xxx shows where incision may be made if lifting cuticle doesn't lead to sufficient drainage.

Fig. 9-10. Draining a paronychia. Perform a digital block as shown in Figure 9-9. After prepping the area, use a scalpel blade to lift the cuticle where infection exists. Sometimes it will be necessary to make a skin incision as well. Place a small strip of iodoform gauze under the cuticle and apply a dressing. Advise the patient to soak the finger in warm water for about thirty minutes every few hours. Also perscribe a course of antistaphylococcal antibiotics.

10

The Chest

Chest Pain 109
    Evaluation 109
    Musculoskeletal Chest Pain 109
    Sternoclavicular Joint Pain 110
Chest Trauma 110
    Rib Bruise or Fracture 111

CHEST PAIN

Evaluation

Note: Although musculoskeletal chest pain is discussed below, far more important is ruling out much more serious and sometimes life-threatening pathology, including conditions of the cardiac, respiratory, and gastrointestinal systems. The differential diagnosis of chest pain is beyond the scope of this book; suffice it to say that no complaint of chest pain should ever be taken to be musculoskeletal in origin until other etiologies have been absolutely excluded by history, examination, and whatever tests are deemed appropriate.

Musculoskeletal Chest Pain (Costochondritis, Pulled Chest Muscle) *(Very Common)*

Diagnosis

    The pain may be intermittent or constant, the duration from hours to years.
    There is usually a history of the pain being affected by movement or by inspiration, though these features may be present with intrathoracic disease as well.

The patient may sometimes give a history of trauma or strain, or of relation to emotional stress.

Examination may reveal tenderness, often over the costal cartilages. This is not an invariable finding since deeper tissues may be involved.

History, examination, and whatever laboratory tests are thought appropriate must be done to rule out cardiac, pulmonary, or gastroesophageal conditions.

In older patients, X-rays or even a bone scan should be done to rule out bony lesions in the ribs.

Pathophysiology

Trauma, strain or overuse of the muscles from direct injury, carrying heavy objects, or coughing and sneezing may be responsible for muscular pain or for pain in the muscular insertions in the costal cartilages.

Stress can lead to chest muscle tightness just as it can to neckache or headache; often worry about chest pain creates a vicious cycle.

Treatment

Reassurance is sometimes all that is needed.

The application of *wet heat* or *ice* may be helpful.

Oral *antiinflammatory medications* can be prescribed.

In cases where psychological factors seem contributory, counselling and/or training in relaxation techniques may be of benefit.

Sternoclavicular Joint Pain *(Uncommon)*

Diagnosis

Tenderness is found over the joint (Figure 10-1).

There will be pain if the shoulder on the affected side is elevated more than one hundred degrees.

Pathophysiology

The joint may be strained acutely or can be painful secondary to degenerative change.

Treatment

Advise *heat* or *ice,* and *antiinflammatory medication* as needed.

CHEST TRAUMA

Note: Open chest trauma or severe blunt trauma is beyond the scope of this discussion; the reader is referred to a text on emergency medicine.

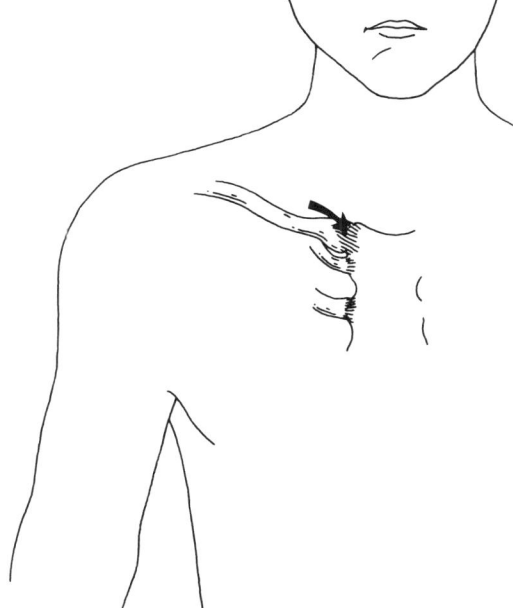

Fig. 10-1. Sternoclavicular Joint.

Rib Bruise or Fracture (Common)

Diagnosis

There is a history of direct trauma and localized pain.

On examination, tenderness of the involved area is found; full breath sounds and a benign abdominal examination help to rule out pneumothorax or a ruptured viscus.

With a fracture, crepitus, ecchymosis, or palpable deformity may be found.

If there is any question of pneumothorax, a chest X-ray must be done.

If the above complications are ruled out, rib X-rays need to be done only:

If the patient requests one.

To help get an idea of how long disability will last (significantly longer if there is a fracture).

In older patients or in those in whom the force of injury seemed so slight in comparison to symptoms as to raise the question of a pathologic fracture.

Note that negative X-ray does not rule out a costochondral separation.

Treatment

Treatment is the same whether a fracture or separation is present or not.

Sufficient *analgesia* should be supplied so that excessive splinting does not occur, because that can lead to atelectasis and pneumonia. For the same reason the patient should be told to take ten deep breaths an hour and to avoid smoking. This is especially important in older patients or those with underlying lung disease. A rib belt may be used if the above precautions are kept well in mind.

11

The Low Back

Introduction	114
Reminder	114
Evaluation	114
History	117
Examination	118
X-ray	120
Laboratory	124
What Causes Low Back Pain	126
Low Back Pain Syndrome (Discogenic)	129
Postural Low Back Pain	129
Lumbar Dysfunction	129
Lumbar Derangement (Central)	130
Derangement With Kyphosis	130
Derangement With Lateral Shift	131
Derangement With Buttock Pain	131
Derangement With Sciatica	131
Derangement With Mild Weakness	132
Derangement With Significant or Worsening Weakness	132
Derangement With Acute Severe Pain	132
Derangement With Bladder or Bowel Symptoms	133
Prolonged Derangement (any type)	133
Recurrent Derangement	133
Chronic Low Back Pain	133
Facet Joint Syndrome	134
Segmental Instability Syndrome	134
Degenerative Arthritis	134
Other Causes of Low Back Pain and Sciatica	135
Low Back Strain	135

Lumbar Sprain	135
Pyriformis Syndrome	135
Compression Fracture	136
Stress Fracture of the Pars Interarticularis (spondylolysis)	136
Spondylolisthesis	137
Ankylosing Spondylitis and Sacroileitis	137
Osteomyelitis and Discitis	138
Neoplasms	139
Fractures and Dislocations	140

Back pain in *children* and *adolescents,* as well as *abnormal curvature* of the back (including scoliosis), are discussed in Chapter 19. *Neurologic symptoms* and *diffuse pain* in the *lower extremities* are the subjects of Chapter 13. Further details on therapuetic suggestions made in this chapter can be found in Chapter 3.

For the anatomy of the lumbosacral spine see Figure 11-1.

REMINDER

Always keep in mind that low back pain can be *referred* from other structures such as the kidneys, prostate, female pelvic organs, gastrointestinal tract, pancreas and abdominal aortic aneurysms. The differential diagnosis of these conditions is beyond the scope of this book, but their possibility must not be forgotten, and appropriate history, physical examination and laboratory tests must be done to rule them out, especially if the clinical picture is not typical for musculoskeletal pain.

ANATOMY

For the anatomy of the lumbosacral spine see Figure 11-1.

EVALUATION

Most low back pain is a result of postural stresses and acute injuries leading over the years to deterioration of the various structures and to acute and chronic pain syndromes. While there still isn't universal agreement regarding the underlying cause and the pathophysiology of this process, a consensus is beginning to emerge that stresses on, and degeneration of, the intervertebral disc plays a central role. (See the box below.)

In evaluating the patient with low back pain, the task of the physician is two fold:

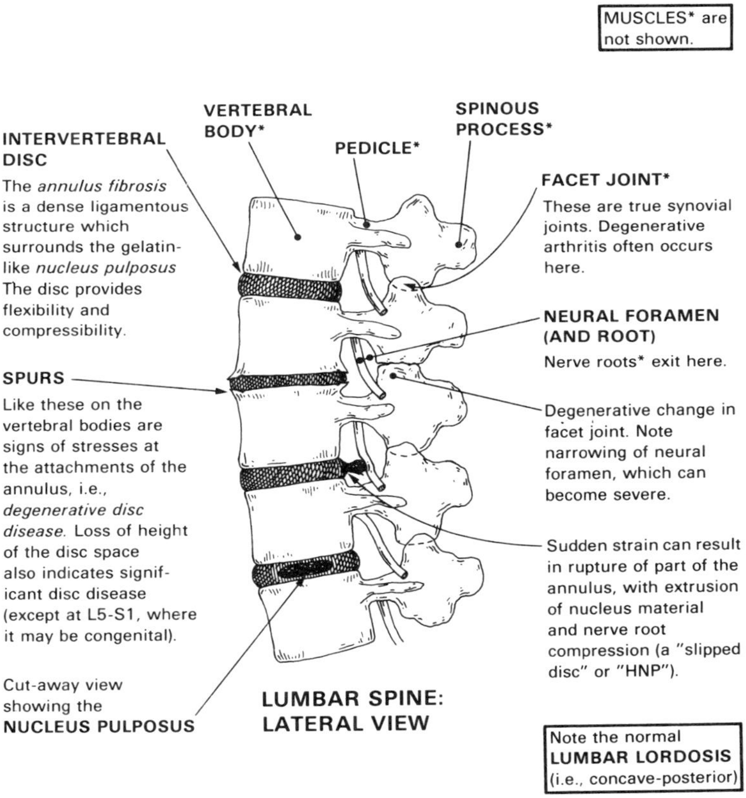

MUSCLES* are not shown.

INTERVERTEBRAL DISC
The *annulus fibrosis* is a dense ligamentous structure which surrounds the gelatin-like *nucleus pulposus*. The disc provides flexibility and compressibility.

VERTEBRAL BODY*

PEDICLE*

SPINOUS PROCESS*

FACET JOINT*
These are true synovial joints. Degenerative arthritis often occurs here.

NEURAL FORAMEN (AND ROOT)
Nerve roots* exit here.

Degenerative change in facet joint. Note narrowing of neural foramen, which can become severe.

SPURS
Like these on the vertebral bodies are signs of stresses at the attachments of the annulus, i.e., *degenerative disc disease*. Loss of height of the disc space also indicates significant disc disease (except at L5-S1, where it may be congenital).

Sudden strain can result in rupture of part of the annulus, with extrusion of nucleus material and nerve root compression (a "slipped disc" or "HNP").

Cut-away view showing the **NUCLEUS PULPOSUS**

LUMBAR SPINE: LATERAL VIEW

Note the normal **LUMBAR LORDOSIS** (i.e., concave-posterior)

Fig. 11-1. Anatomy of the lumbosacral spine. Asterisks indicate pain-sensitive structures.

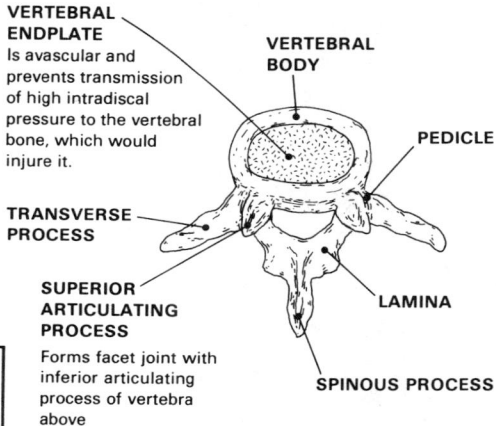

VERTEBRAL ENDPLATE
Is avascular and prevents transmission of high intradiscal pressure to the vertebral bone, which would injure it.

VERTEBRAL BODY

PEDICLE

TRANSVERSE PROCESS

SUPERIOR ARTICULATING PROCESS
Forms facet joint with inferior articulating process of vertebra above

LAMINA

SPINOUS PROCESS

A LUMBAR VERTEBRA* TOP VIEW

Degenerative changes of the annulus fibrosis (secondary to mechanical stresses), along with loss of fluid content of the nucleus, lead to small malalignments between vertebrae; postural abnormalities lead to malalignments as well. These malalignments result in stress on the facet joints, with the subsequent development of degenerative arthritis.

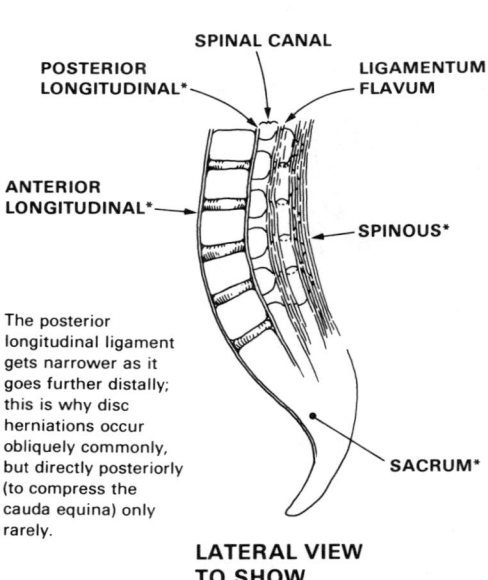

POSTERIOR LONGITUDINAL*

SPINAL CANAL

LIGAMENTUM FLAVUM

ANTERIOR LONGITUDINAL*

SPINOUS*

The posterior longitudinal ligament gets narrower as it goes further distally; this is why disc herniations occur obliquely commonly, but directly posteriorly (to compress the cauda equina) only rarely.

SACRUM*

LATERAL VIEW TO SHOW LIGAMENTS

Fig. 11-1. Continued.

HISTORY

1. To rule out referred pain from other structures, or the less common causes of musculoskeletal pain such as compression fractures, tumors, sacroileitis, etc.
2. Once these diagnoses have been ruled out by history, examination and/or whatever tests are deemed necessary, and it has been decided that the patient indeed suffers from the usual "low back pain syndrome," it is necessary to determine how far along in that process the patient's back has gone and what paths it has taken, so that a choice may be made from among the many treatment methods available.

If these two facets of evaluation are kept mentally separate and taken in order, it is much less likely that serious pathology will be missed and more likely that a good understanding of the patient's problem will be obtained and appropriate therapy instituted.

HISTORY

Ask about the *location* of the pain. Discogenic low back pain is usually felt centrally or bilaterally, often with referred pain and tenderness in one sciatic notch area or down the leg. Pain localized to one side higher up than that may be due to a MUSCLE STRAIN (with a history of sudden onset and palpable tenderness), to a FACET JOINT SYNDROME (in an older patient, with the pain usually made worse by extension) or to a STRESS FRACTURE OF THE PARS INTERARTICULARIS (unilateral belt line tenderness in a teenage athlete whose sport involves repetitive extension, e.g., gymnastics).

Was there any specific *precipitating trauma or motion*? DISC DERANGEMENTS may or may not be associated with such an onset; strains almost always are. High energy indirect trauma (such as motor vehicle accident) with subsequent diffuse pain is consistent with a LUMBAR SPRAIN or a COMPRESSION FRACTURE.

When does the pain occur? Pain which is there only when certain prolonged postures are maintained (sitting, standing, or on first awakening in the morning) is classified as POSTURAL LOW BACK PAIN. Momentary mild pain at the end-range of normal motion is consistent with loss of flexibility, referred to as DYSFUNCTION; while severe pain only with movement in a patient with a long history of significant back problems may indicate SEGMENTAL INSTABILITY due to a severely degenerated disc. (Of course during an acute disc derangement, motion and certain postures will exacerbate the pain too). Pain occurring mainly at night and not definitely affected by motion is very suspicious for metastatic or primary MALIGNANCY, even more so in older patients. (New onset of low back pain or a change in the character of the symptoms in patients over 50 should also be considered malignancy until proven otherwise.)

Inquire as to the *duration* of symptoms and ask about *previous episodes* (discogenic low back pain and sciatica is usually a chronic recurrent disease), as well as previous diagnoses, X-rays, and treatments.

Is there *radiation* of pain into the legs? Radiation past the knee is pathognomonic of nerve root impingement by a disc (or in older patients, sometimes by hypertrophic changes). Radiation to the anterior thigh is typical of L4 root lesions; the lateral thigh, anterior shin and dorsal foot, L5; the posterior thigh and calf and sole of the foot, S1. A vague radiation into both legs, if it occurs mainly with walking, is suggestive of the PSEUDOCLAUDICATION SYNDROME; but bilateral symptoms should always bring to mind the possibility of spinal cord tumors, especially with bilateral neurologic findings (though indeed often discs are found to be responsible). Ask about *paresthesias, numbness,* and *weakness* in the legs.

Psychosocial issues often become prominent in prolonged and chronic cases. Be sure to evaluate the patient's psychologic status, and how his condition is affecting his life; consider loss of income, secondary gain, fear of disability, psychogenic muscle tension and depression.

And finally, be sure to ask about gynecologic, urinary, gastrointestinal, and systemic symptoms, to help rule out nonmusculoskeletal or infectious causes of the patient's symptoms, or at least to indicate if more work-up is necessary.

Examination

Inspect the back for deformity. (See Figure 11-2). An especially prominent spinous process may be indicative of SPONDYLOLISTHESIS. If the patient is bent over forward and can't straighten up he probably has a large central disc derangement. A lumbar scoliosis can be due to structural scoliosis (discussed in Chapter 19), but more commonly is associated with acute disc derangements in which the bulge is posterolateral. (In these cases extension will be painful if performed before this "lateral shift" is corrected.) Also look for uneven iliac crests, by palpating them simultaneously (beware the crooked underwear line!); this may indicate leg length discrepancy, which may be a contributing factor to low back pain (see Figure 11-3).

Look for *tenderness.* This can be found over the involved muscle in a strain; usually none will be found in a central derangement. If there is radiation of the pain to the buttock or leg, tenderness over the sciatic notch is common. Tenderness over the SI joints is often a sign of inflammatory SPONDYLITIS.

Check *range of motion.* With a derangement, pain with flexion is common, and repeated flexion may worsen or peripheralize the pain (i.e., make it radiate further). Also look for deviation to one side with forward flexion, indicating a posterolateral disc bulge. Extension while standing, especially if repeated, will often alleviate the pain of a disc derangement; if it doesn't, try extension in the prone position. At least the pain should begin to centralize. Early in the course of disc derangements, while there still is a significant inflammatory response to the

HISTORY

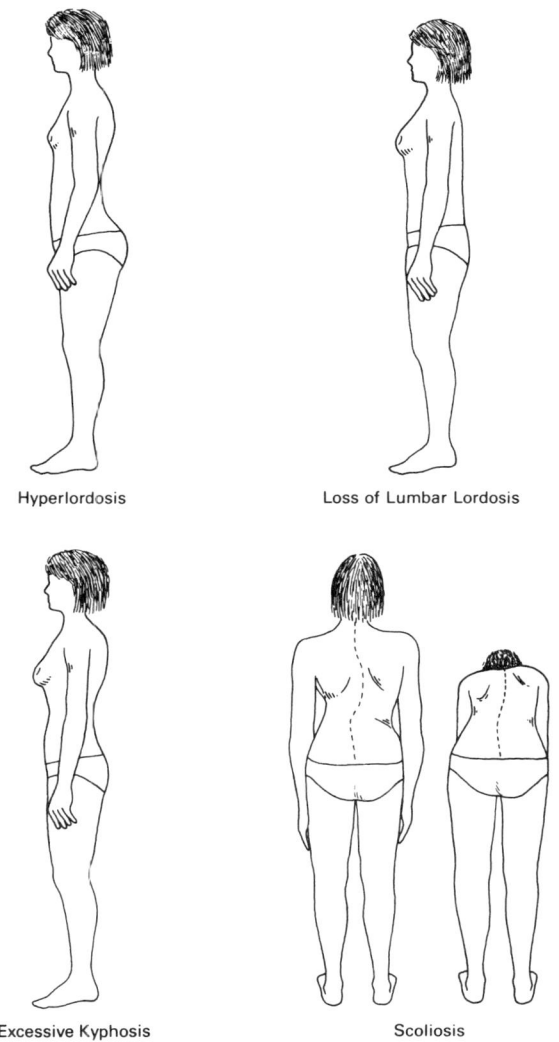

Fig. 11-2. Postural abnormalities.

disc bulge, extension may be irritating. If it is, look again for a subtle lateral shift. Localized unilateral paralumbar pain with extension may indicate a facet syndrome. Restricted flexion in a patient with chronic low back ache may be a clue pointing toward SPONDYLITIS or may be due simply to deconditioning and loss of musculoligamentous flexibility. This last condition more commonly affects the range of *extension* motion, and is called DYSFUNCTION which is discussed below. Two tips:

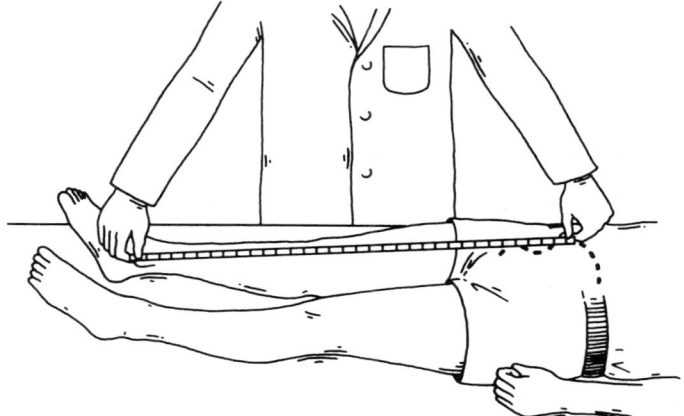

Fig. 11-3. Measuring leg length. Measure from the ASIS to the medial malleolus. Even this can be inaccurate; an X-ray called a *scanogram* should be ordered if determining leg lengths is thought to be important.

1. Flexion range can be quantified and followed by seeing how far the patient's fingers can come to the floor (as lateral range of motion can be evaluated by seeing how far down the leg the fingers go).
2. When checking extension range, be sure the patient is arching his back, not flexing his knees.

If there is pain into the buttock or leg, also examine the hip (Figure 11-4), perform a straight leg raise test (Figure 11-5), and examine vascular status (pulses, capillary filling, skin color and temperature) and neurologic status (strength and pin sensation) in the legs (see Figure 11-6).

If tenderness was found over the SI joints or spondylitis is suspected for other reasons, the SI joints can be tested as in Figure 11-7.

If there is any suggestion at all that the pain may not be musculoskeletal in origin, be sure to do an adequate abdominal examination (including checking for bruits), punch the CVA's, looking for tenderness, and do a rectal and a pelvic or prostate examination.

X-Ray

Although when X-rays should be taken was discussed in Chapter 2, the situation is somewhat more complicated in the case of the low back for two contradictory reasons. On the one hand, because of the amount of soft tissue that has to be tranversed, low back X-rays involve *more radiation exposure,* and the exposure is to a variety of particularly radiosensitive tissues (the gonads and the kidneys included). On the other hand, certain *very serious conditions* such as

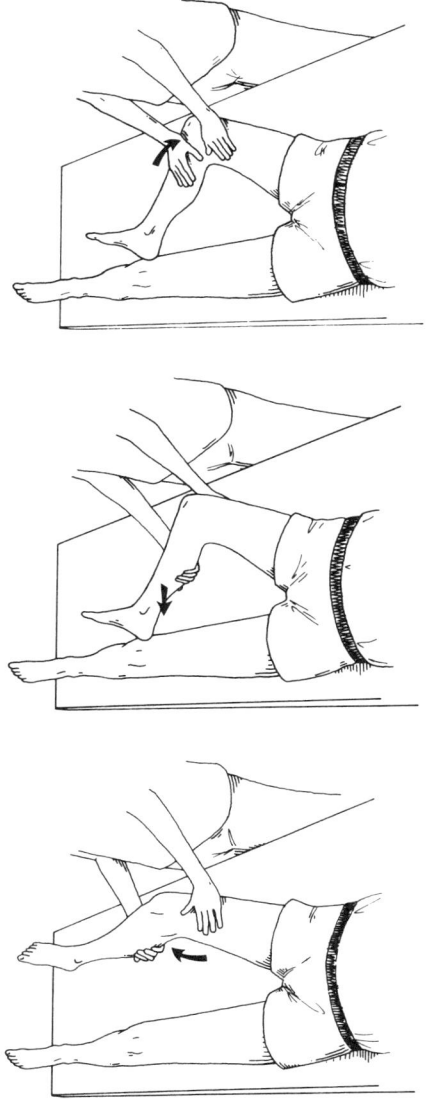

Fig. 11-4. Patrick's Test to rule out hip pathology.

osteomyelitis or tumor will be easily missed for a significant period of time if the criteria for taking X-rays are too stringent, and other lesions that may affect the treatment course of mechanical low back pain will not be diagnosed. Therefore, the following criteria for X-raying the lumbosacral spine are suggested:
- A history of direct trauma.
- A history of indirect trauma with immediate severe pain.
- *Any* objective neurologic findings.
- Fever, chills or sweats without other obvious source.
- History or examination suggestive of sacroileitis (for example, a family history of ANKYLOSING SPONDYLITIS, a personal history of psoriasis or colitis, or positive sacroiliac joint tests. In such cases order sacroiliac joint views).
- Examination suggestive of spondylolisthesis.
- Any patient under 25 without obvious history of strain or with pain persisting for more than two weeks.
- Any patient over 50 or with a past history of malignancy.
- Any case in which pain is prolonged (i.e., does not resolve as expected).
- A patient who needs the reassurance of a normal X-ray.

A *bone scan* should be considered the next step in cases in which the patient's age, past history, and current symptoms or examination are suggestive of the possibility of infection, inflammation or neoplasm and in whom X-rays are negative.

Note that the following X-ray findings, though thought to be etiologic factors of low back pain in the past, are now considered to be no more than incidental findings in most cases, though some predisposition to degenerative disc disease is possible:

        Spina bifida
        Hemivertebra
        Iliotransverse joint
        Lumbarization of S1
        Sacralization of L5

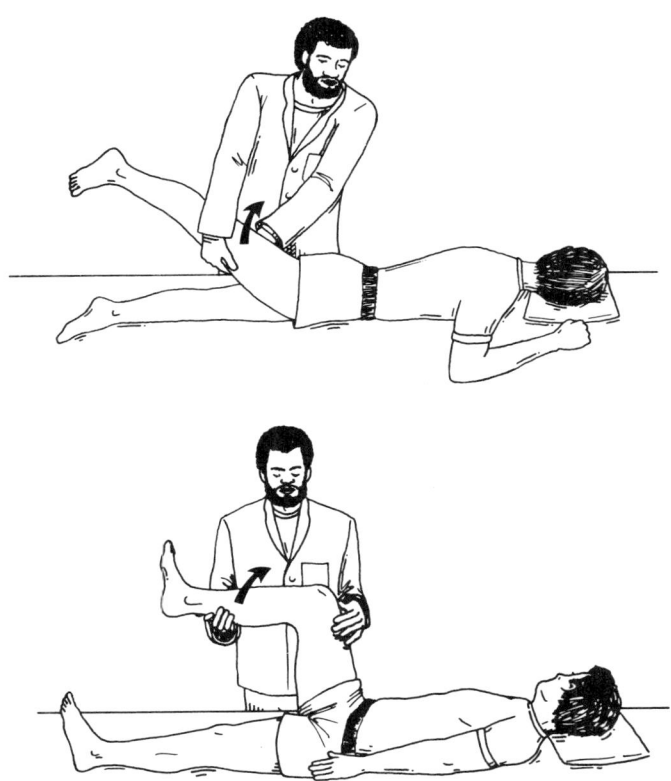

STRAIGHT LEG RAISING:
This is positive if it produces pain in the leg, or if *LeSegue's Modification* is positive. If the patient complains of back pain as the leg is being straightened, flex it a little bit to remove the pain and then see if the pain recurs with forced dorsiflexion of the ankle.

Fig. 11-5. Radicular signs. (top) Reverse Straight Leg Raising. (bottom) Straight Leg Raising. These are positive if they produce pain in the leg, or if *Le Segue's Modification* is positive. If the patient complains of back pain as the leg is being straightened, flex it a little bit to remove the pain and then see if the pain recurs with forced dorsiflexion of the ankle.

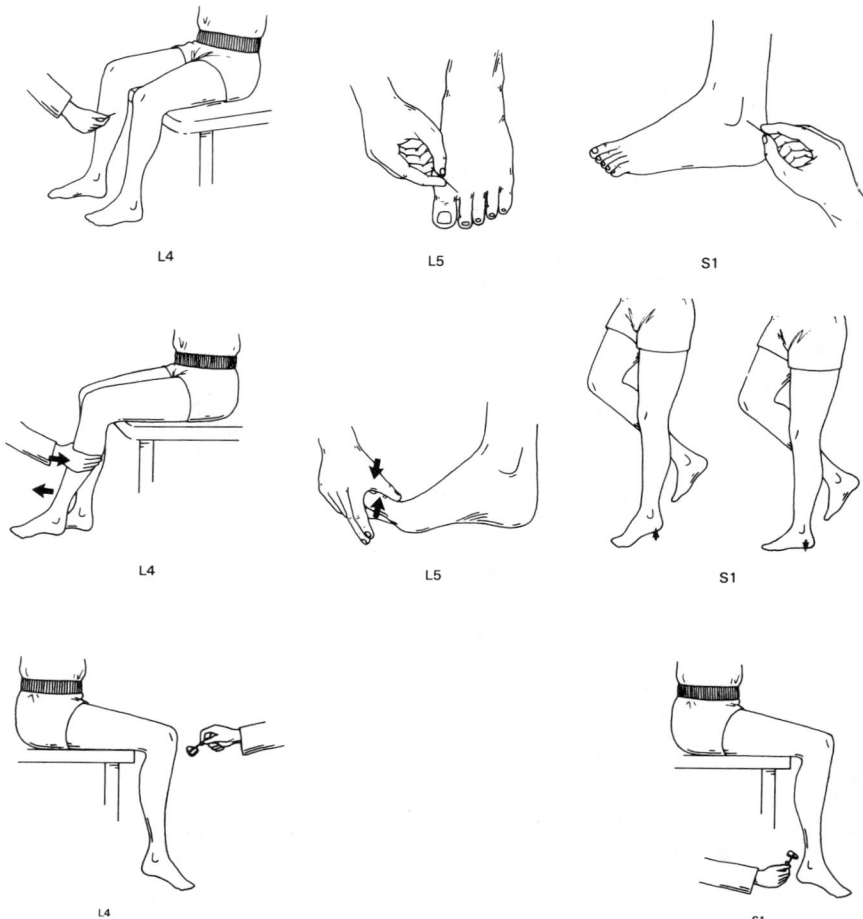

Fig. 11-6. Screening lower extremity neurologic examination in patients with low back pain.

Other Lab Tests

See the discussion in Chapter 2. Note, though, that the *ESR* is an extremely useful test in differentiating mechanical lower back pain from that arising from primarily inflammatory lesions such as infection or spondylitis.

HISTORY

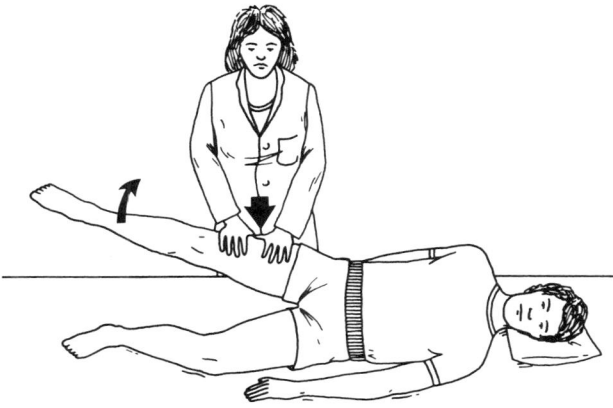

RESISTED HIP ABDUCTION:
Side with the suspected SI joint pathology is up.

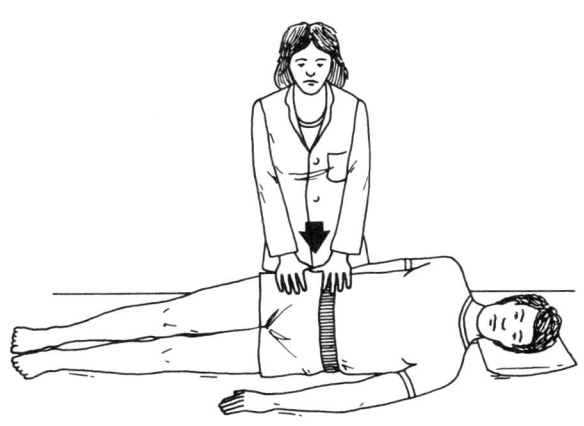

SI JOINT COMPRESSION
Fig. 11-7. Sacroiliac joint signs.

Of course, a *urinalysis* will often be necessary to rule out a urinary tract cause for the patient's pain, especially in females.

The HLA-B27 marker has a much higher incidence in patients with spondylitis.

Other lab tests used, especially in older patients, to rule out tumor are CBC, calcium, phosphate, alkaline phosphatase, acid phosphatase, and serum protein electrophoresis.

What Causes Low Back Pain?

Nobody knows.

Obviously pyelonephritis, leaking aortic aneurysms and other serious internal conditions can cause pain referred to the back. And we all know that stress fractures, compression fractures, bone infections, tumors, and inflammatory spondylitis are sometimes responsible. But in the overwhelming majority of people complaining of low back pain none of these are found. They have "strains" or "derangements" or "slipped discs" or "facet syndromes." They have back pain that is central or unilateral, that radiates to the buttock or all the way to the foot or not at all, that is acute or recurrent or chronic, mild or severe. Do these people really have a multitude of different pathologies? Or perhaps does one of the dozens of theories about low back pain that have been proposed over the past decades explain it all? Which one?

Nothing has been proven yet, of course, but lately a consensus has begun to emerge that stress on and subsequent deterioration of the *intervertebral disc* is at the center of a process that affects almost all of us to one degree or another and is responsible not for only severe sciatica with neurologic signs, but indeed for most of the low back suffering in the world. This model is internally consistently and fits what is observed in actual patients over time better than previous concepts. Before describing it, there follows a short discussion of two other theories which currently affect, sometimes almost subliminally, much of the treatment that is prescribed.

First, the ubiquitous "low back strain." A strain is of course a muscular and/or tendinous injury. This is often said to be associated with "muscle spasm" and is named as the cause of the patient's discomfort. But only rarely can specific muscle tenderness or spasm be palpated in the lumbar area in patients complaining of low back pain.* And as an explanation for the occasional deformities that accompany acute low back pain, spasm just doesn't hold up, so to speak. Consider: the patient who comes in bent over forward with low back pain is said to have muscle spasm. But which muscle? The paraspinal muscles are *extensors*, not flexors. What muscles are the primary flexors of the back? The abdominals and iliopsoas. Do people who come in bent over with low back pain have abdominal rigidity or flexion contractures of the hips? Of course not. It's not muscle spasm, it's a protective posture which the patient assumes to avoid pinching off the bulging disc.

Strains of the paralumbar muscles and spasm there *do* occur occasionally (especially in athletes) and are easy to diagnose because of localized tenderness and/or spasm. (A patient who has had a strain and a disc derangement at different times can easily tell them apart.) Strains and spasms however are not adequate explanations for the great majority of acute and chronic low back pain.

Another explanation for common low back pain that has been widely accepted for many years is that of excessive lumbar lordosis. Two generations of back suf-

---

*This is in contrast to patients complaining of neck or upper back pain, where palpable tenderness and spasm are commonly found.

ferers have been told that what is necessary to keep a back healthy or cure pain that is already there is to maintain a flat-back posture, and they have faithfully performed a series of exercises (Williams' exercises, named for the physician who designed them*) to reduce their lordosis, all without any good evidence that it is helpful, or even a good theory as to why it should be. Indeed, like the muscle spasm theory discussed above, the excessive lordosis theory doesn't fit the facts as seen in most patients. Almost always low back pain is worse after sitting, despite the fact that lordosis is minimized or reversed during sitting; and acute episodes are often brought on by bending forward, again reversing the lordosis. Further, only a minority of low back sufferers are found to have excessive lordosis (usually obese women).

But the greatest theoretical difficulty with the models described above and many others is that they separate out syndromes (strain versus slipped disc, e.g.) which are observed over time to blend into one another or progress from one to another in the same patient, sometimes during a single episode.

A Unified Model

The concept of the disc as the source of most low back discomfort was first popularized by Cyriax, and later expanded upon and more closely correlated to the common clinical syndromes by Robin McKenzie, a physical therapist from New Zealand. Their ideas can be fitted into a cogent model of the low back pain syndrome that is internally consistent and in keeping with what is seen in actual patients. It is helpful in understanding what one sees in individual patients to understand the low back syndrome as a *process*, to wit:

1. Postural stress, especially if prolonged, leads to migration of the gelatinous nucleus pulposus within the disc, and subsequent reversible deformities of the annulus fibrosis. There may also be strain on other musculoligamentous structures, especially if there is insufficient flexibility. (McKenzie calls this state of insufficient flexibility in certain directions *dysfunction*). *By far* the most common postural stress is *prolonged sitting*. During sitting the disc is forced posteriorly by the flattened or reversed lumbar lordosis. (See Figure 11-8). The bulging disc can exert pressure on pain-sensitive structures.
2. Acute trauma, major (as in a sudden jerk with heavy or improper lifting) or more commonly minor sudden movements, and usually bending forward and especially if performed after prolonged sitting or other abnormal posturing, may lead to a sudden disc bulge and acute pain. Straight posterior bulging results in central or bilateral low back pain; if the protrusion is to one side, as it often is because of the restraining influence of the posterior longitudinal ligament in the center, the pain is unilateral, and may radiate to the buttock or down the leg. If nerve root impingement is more pronounced, the pain will radiate fur-

---

*Recently this series of exercises has at least been modified to exclude sit-ups, which have been shown to markedly increase intradiscal pressure.

ther and may be associated with numbness, paresthesias, or sometimes true weakness. If there is a large protrusion the patient may assume a protective posture: bent over forward (for a posterior protrusion), or scoliotic (for a posterolateral protrusion).

3. With recurrent derangements and/or prolonged postural stresses, the substance of the annulus fibrosis deteriorates, thus making further derangements more likely, harder to treat and more prolonged.
4. Chronic degeneration of the disc results in instability, which can be "macroscopic" (leading to a movement-related pain syndrome) or only "microscopic," allowing stresses to increase at the posterior joints between the vertebrae and thus accelerating degeneration of these facet joints (i.e., DJD). Hypertrophic spurring associated with this joint degeneration may itself impinge on the nerve roots and cause leg pain or neurologic symptoms.
5. The combination of bulging or ruptured discs, hypertrophic changes in the spine and a congenitally narrowed spinal canal can sometimes result in spinal stenosis and pseudoclaudication (discussed in Chapter 13).
6. A chronically degenerated disc (or one to which extreme forces are applied, as in a motor vehicle accident) can actually *rupture* (herniate), often causing radicular symptoms not clearly affected by position and less likely to resolve spontaneously than a simple derangement. A large rupture can lead to especially severe pain because of the inflammatory effect of the extruded nucleus. Rarely a large midline protrusion can compress the cauda equina, resulting in bowel and bladder dysfunction along with neurologic signs in the legs.
7. Recurrent pain may become more continuous, and patients whose psyche or life situation predisposes them to it can develop a chronic pain syndrome.

So in summary, according to this model most acute and chronic low back pain and radiated leg pain is the result of a process by which the intervertebral discs (most commonly L5-S1, also L4-5 and to a lesser extent higher levels) are affected by postural stresses and trauma, bulge, deteriorate, lead to degenerative change in the facet joints and sometimes rupture.

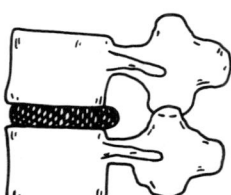

Fig. 11-8. Spondylolisthesis and spondylolysis.

HISTORY

Low Back Pain Syndrome *(Extremely Common)*

Understand that the *process* of postural/degenerative low back pain is divided into the following sub-classifications not because there are clear dividing lines between them, but because these categories help in making management decisions. Keep in mind also when evaluating the effectiveness of various treatments that most attacks of low back pain and sciatica are self-limited; prevention of further injury and symptomatic relief are probably the most important therapeutic goals. Appendix 11-1 lists commonly used treatments for the low back pain syndrome and when they are most likely to be useful.

Postural Low Back Pain

Diagnosis

Low back pain (and often pain in the upper back and neck as well) is brought on by certain positions, especially if prolonged, and relieved with movement. Sitting is the most common precipitant.

Pathophysiology

Poor posture places stresses on the disc and other musculoligamentous structures.

Treatment

Instruction in proper posture (Patient Handout 11-2).

Lumbar Dysfunction

Diagnosis

Pain occurs at the end-range of movement. Often a contributing factor to postural low back pain and disc derangements, since the loss of range of motion precludes proper posture and concentrates stress on the disc.

Pathophysiology

Loss of flexibility, because of poor posture and/or scarring from previous injuries, means that structures are placed under stretch prematurely.

Treatment

Postural instruction and exercise to restore flexibility. The patient must understand that the latter often temporarily increases pain because shortened structures are being stretched to restore their proper function. Evaluation and follow-up by a physical therapist is best.

Lumbar Derangement *(central)*

Diagnosis

The pain may or may not come on with specific trauma or a specific movement. It is felt in the center of the low back or across both sides. The pain does not radiate. Pain is usually worse after sitting and better with movement or recumbency.

On examination, there is usually no significant tenderness found and no deformity noted. The pain is made worse by flexion; extension will usually make it better or have no effect, though with more severe cases early in the course extension may also increase the discomfort.

Pathophysiology

A midline posterior disc bulge.

Treatment

Ice applications in twenty minute sessions may be helpful.
Advise *aspirin* or prescribe *NSAI'S*.

Instruct the patient in avoiding bending, lifting and twisting motions as well as sitting (especially prolonged sitting) while symptomatic. If any sitting is done it should be with proper lumbar support and frequently interrupted. Teach the patient extension exercises. See Patient Handout 11-1. Teach the patient, or refer to a physical therapist or back school for, proper ergonomics, posture and exercise (especially to correct dysfunction) and self care instruction. Prescribe pain medication as necessary.

N.B.:

In the discussion of the following variations and sequelae of disc derangements, only differences from a central disc derangement are mentioned. Please refer back to the discussion of that entity immediately above for basic information and treatment.

Derangement with Kyphotic *(bent-over)* Deformity

Pathophysiology

The patient assumes the bent over posture to protect the protruding disc.

Treatment

Relief can often be obtained rather quickly in the following manner: have the patient lie down prone on the examination table with enough pillows under his abdomen to be comfortable. Remove a pillow and allow enough time for the patient to become comfortable again (usually ten to fifteen minutes), then remove another. Repeat the procedure until the deformity is corrected; then begin extension exercises in the prone position. Continue treatment as for an uncomplicated central derangement.

Derangement with Scoliotic *("lateral shift")* Deformity

Pathophysiology

A protective posture for a posterolateral disc protrusion.

Treatment

Before extension can be accomplished without pain, the lateral shift must be corrected. Have the patient glide hips and shoulders in opposite directions without tilting either (like giving a "hip check") and then extend. Teach him how to do this at home. Sometimes it is difficult to hold the correction. A properly trained physical therapist can be very helpful in this situation.

Derangement with Pain Radiated to Buttock

Diagnosis

Sciatic notch tenderness is almost invariable. Be sure to check the hip carefully.

Treatment

As for a central derangement. Be sure the pain is centralizing, not peripheralizing, with the exercises.

Derangements with Pain Down the Leg (sciatica)

Diagnosis

The patient often reports that walking makes the pain worse. Be sure to check neurovascular status carefully in the leg. A straight leg raising test can be used to follow progress.

Pathophysiology

Nerve root irritation by the disc. (Note that sciatica can also be due to nerve root impingement by degenerative change in the bony structure (degenerative arthritis), or sometimes by a tight pyriformis muscle (pyriformis syndrome).

Treatment

As for a central derangement. Be sure the pain is centralizing, not peripheralizing, with the exercises.

Derangement with Mild Weakness*

Pathophysiology

More nerve root impingement and possibly damage.

Treatment

Insure that the patient understands well the need to follow instructions, especially regarding precautions against reinjury, lest the problem worsen. Follow the patient very closely.

Derangement with Significant or Worsening Weakness

Pathophysiology

A true disc herniation is more likely. Consider the possibility of spinal tumor.

Treatment

Refer to an orthopedist or neurosurgeon. A course of high-dose rapidly tapering steroids and bed rest is warranted. The patient needs a CT scan and probably surgery.

Derangement with Acute Severe Pain

Diagnosis

Be especially careful to rule out other causes, such as leaking abdominal aortic aneurysm, kidney stone, fracture or others.

Pathophysiology

A large inflammatory response from an acutely ruptured disc.

Treatment

Prescribe bed rest and high-dose rapidly tapering steroids. Be sure to supply adequate pain medication. Consider hospitalization and a trial of pelvic traction.

---

*The presence of paresthesias or numbness generally does not affect management.

Derangement with Bowel or Bladder Symptoms

Pathophysiology

Compression of the cauda equina.

Treatment

Start high dose steroids and obtain STAT neurosurgical consultation.

Prolonged Derangement *(any type)*

Diagnosis

Take another look to rule out other causes. Look for trigger points which may respond to local injection. Check for dysfunction, and for compliance with prescribed precautions and treatments.

Treatment

Refer for consideration of traction, other physical therapy modalities, or manipulation. If sciatica is present consider referral for epidural steroid injection. Perform a CT scan. Consider referral for surgery or chymopapain if all else fails after a few months.

Recurrent Derangements

Treatment

Advise weight loss if indicated, strengthening exercises especially for the abdominals, range of motion exercises, and strongly consider referral to a back school.

Chronic Low Back Pain

Diagnosis

Be especially careful to rule out other causes, especially spondylitis. Evaluate the patients psychologic status and life situation. Consider if a "chronic pain syndrome" has developed.

Treatment

The following may be helpful: antidepressant medications; an aerobic exercise program; vocational rehabilitation; behavior modification techniques; trigger point injection; TENS unit; referral for psychosocial evaluation.

Facet Joint Syndrome

Diagnosis

The pain is usually felt unilaterally but close to the spine, and is made worse by extension. More common in older patients.

Pathophysiology

Pain originates in a degenerated, inflamed facet joint.

Treatment

Steroid injection into the area of the facet joint may be helpful. If this or the injection of a local anesthetic is temporarily helpful, consider referral for facet rhizotomy surgery (where the nerves leading from the inflamed joint are cut).

Segmental Instability Syndrome

Diagnosis

Moderate to severe pain in the spine occurs with certain motions. X-rays in flexion and in extension ("dynamogram") can reveal excessive motion between two vertebrae.

Pathophysiology

Severe disc and soft tissue damage or deterioration.

Treatment

Strengthening exercises and ergonomic instruction by a physical therapist is critical. A fitted corset can be helpful. Otherwise consider referring the patient for evaluation for fusion surgery.

Degenerative Arthritis

Diagnosis

By X-ray.

Pathophysiology

Simply a result of the degenerative process that has been discussed above. Chronic sciatica can be due to nerve root impingement by a hypertrophied joint rather than by a disc.

Treatment

Postural and ergonomic instruction, ice or heat, aspirin or prescription NSAI's, and specific treatment for specific features as noted above.

OTHER CAUSES OF LOW BACK PAIN AND SCIATICA

Low Back Strain *(Common)*

Diagnosis

A specific paraspinal muscle is strained by abnormal posturing or sudden movement. On examination, localized tenderness is found.

Treatment

Advise the patient to protect from further injury and adapt proper posture as for a lumbar derangement. Apply ice for twenty minutes at a time. Use aspirin and NSAI, a "muscle relaxant," or pain medications as necessary.

Lumbar Sprain *(Uncommon)*

Diagnosis

This is a term used to denote the diffuse lumbar pain which can occur following high-energy injuries such as in a motor vehicle accident. X-rays are normal.

Pathophysiology

Diffuse soft tissue injury; disc injury can coexist.

Treatment

As for a strain. The course is often prolonged, and referral to a physical therapist is recommended.

Pyriformis Syndrome *(Uncommon)*

Diagnosis

The patient complains of sciatic pain, without objective neurologic findings, which is usually worse with walking. It is probably especially common in athletes.

On examination pain will be reproduced by resisting internal rotation of the hip with the patient sitting.

Pathophysiology

The sciatic nerve is irritated by the tight pyriformis muscle. How often this is actually a cause of sciatica is still widely debated.

11. THE LOW BACK

Treatment

A pyriformis stretching program supervised by a physical therapist may be helpful.

Compression Fracture *(Common in older patients)*

Diagnosis

The patient can usually well localize the origin of the pain in the spine. The onset is usually sudden, with a certain movement, sneeze, etc. It is most common in older patients with osteoporosis, but can occur in younger patients with more severe trauma. Always consider the possibility of tumor involving the vertebra. Examine carefully for neurologic damage. Watch for a few days of ileus.

Treatment

Rest, prevention of further injury, bracing, and pain medication as necessary. Treat osteoporosis as indicated (see Chapter 26).

Stress Fracture of the Pars Interarticularis
(spondylolysis) *(Common only in young
athletes in activities requiring repetitive extension)*

Diagnosis

The pain is unilateral at the belt line, without radiation, and is made worse by extension. The syndrome is common in gymnasts, ballet dancers and football linemen. On examination, having the patient stand on one leg (on the affected side) and arch her back will reproduce the pain.

X-ray sometimes shows spondylolysis, especially on the oblique view, but this may be residual of a developmental ossification defect, i.e., not related to current symptoms; conversely, X-ray is often negative early in the course. A bone scan is a sensitive indicator, but not always necessary for diagnosis.

Pathophysiology

A stress fracture.

Treatment

The patient must refrain from the offending motion until symptoms have resolved, often several months. The earlier the diagnosis is made and precipitating motions avoided, the sooner healing occurs. Prolonged bilateral symptoms could lead to a spondylolysthesis.

OTHER CAUSES OF LOW BACK PAIN AND SCIATICA

Spondylolisthesis (Uncommon)

Diagnosis

Forward slippage of a vertebra on the one below it (most commonly L5 on S1) is noted on X-ray (see Figure 11-9).

It can cause low back pain and/or nerve root compromise.

Pathophysiology

Most commonly spondylolisthesis is due to spondylolysis (which itself can be a developmental defect or a stress fracture), but can rarely be due to isthmic stretching in a child, to trauma, or itself be a congenital defect.

Treatment

Many cases are asymptomatic. The chance of a spondylolisthesis seen on X-ray being the etiology of the patient's symptoms is probably greater the younger the patient is.

In spondylolytic spondylolisthesis, methods of treatment are fairly similar to those in other etiologies of low back pain. Flexion exercises (Patient Handout 11-3) are probably of greater importance, however, and patients with fairly servere symptoms may be helped by a properly fitted brace.

The management of the other forms of spondylolisthesis and of moderate to severe cases of spondylolytic spondylolisthesis should be left to the specialist.

Ankylosing Spondylitis and Sacroileitis *(Uncommon)*

Diagnosis

Low back pain occurs in a young male.

Examination shows positive sacroiliac joint tests and tenderness over the SI joints. Limitation of forward flexion and a limitation of chest expansion (because of involvement of the costovertebral joints) occurs as the disease progresses.

X-ray shows sacroiliac joint changes (blurring of the margins, irregular ero-

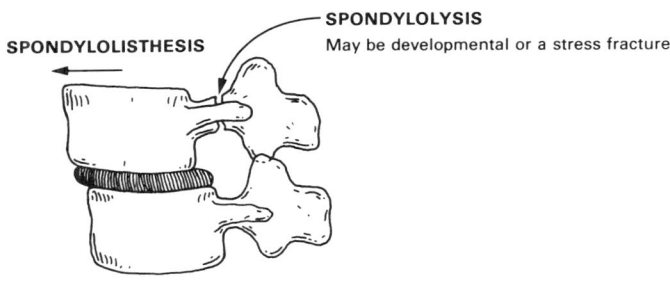

Fig. 11-9. Posterior Disc Bulge.

sions and sclerosis*) and syndesmophytes (distinguished from the osteophyte of degenerative disc disease by the fact that the former extend more vertically).

An elevated ESR and a positive HLA-B27 antigen are found in 90% of cases.

About one-fourth of patients have large peripheral joint involvement, usually of the hips and shoulders.

About one-fourth of patients have iritis at some time in the course of their disease.

A small percentage of patients develop aortic insufficiency.

Similar syndromes may be seen in patients with psoriatic arthritis or the arthritis associated with inflammatory bowel disease. Back symptoms in patients with Reiter's syndrome are usually not significant. See Chapter 23 for further discussion of these conditions.

Treatment

Use *antiinflammatory* medication.

Postural instruction to minimize the eventual deformities by maintaining a straight spine is critical. Also important are breathing exercises to maintain chest expansion.

Referral to a physical therapist for long-term follow-up is wise.

Patients with cardiac or eye involvement should be followed by specialists in those fields.

Osteomyelitis and Discitis *(Rare)*

Diagnosis

To make the diagnosis, the condition must be suspected, since often fever and other constitutional symptoms are absent, and X-ray changes may not become apparent for weeks.

The ESR is invariably elevated, and should be performed in any patient whose course is at all prolonged or otherwise atypical, as well as in high-risk groups (patients with recent pelvic infection or back surgery, IV drug users, children and adolescents with back pain).

A bone scan will be positive before evidence of infection can be seen on X-ray.

---

*Not to be confused with OSTEITIS CONDENSANS ILII, a radiologic finding of no significance, usually in multiparous women, in which sclerosis is confined to the iliac side of the joint. This condition is analogous to the equally insignificant OSTEITIS PUBIS (sclerosis of the pubic symphisis), found in runners, soccer players, and horseback riders as well as mothers.

OTHER CAUSES OF LOW BACK PAIN AND SCIATICA

Pathophysiology

Vertebral osteomyelitis is usually secondary to staph aureus or, in needle users, to gram-negative organisms; the bacteria most commonly reach the spine by hematogenous spread.

Disc space infection is most common in children and teenagers and is thought to be viral (except in postlaminectomy patients).

Treatment

These patients should be *referred* to an orthopedist and an infectious disease specialist for management.

Benign Neoplasms *(Rare)*

These are usually found in patients under 30.

The pain is not necessarily relieved by rest.

The diagnosis is by X-ray and/or bone scan.

These tumors are almost always found in the posterior elements of the vertebra.

Osteoid osteoma is the most common type; others are osteoblastoma, eosinophilic granuloma, aneurysmal bone cyst, and giant cell tumor.

The patient should be referred to an orthopedic surgeon.

Primary Malignant Neoplasms *(Uncommon)*

These usually occur in patients over 40.

Again, pain is not necessarily relieved by rest.

These tumors involve the vertebral bodies.

Myeloma is most common: X-ray may show lytic lesions or just diffuse osteoporosis. Diagnosis is by abnormal serum protein electrophoresis; the ESR is invariably increased but is not specific. Patients with myeloma may have constitutional symptoms as well. See an internal medicine or hematology text for further details.

Chordoma is the only other primary malignant tumor of the spine seen at all commonly.

Patients should of course be referred to appropriate specialists.

Metastatic Tumors *(Common)*

The spine is the most common site of metastatic spread in the skeleton.

Only a minority of vertebrae shown to be involved at autopsy demonstrate lesions on X-ray.

In a fair number of patients with malignancy, back pain secondary to metastasis will be the presenting symptom.

Therefore, *patients over 50* presenting with back pain, *especially* if without known trauma or strain and/or if unrelieved by rest, should be *considered to have a spinal metastasis until proven otherwise.*

If there is any level of suspicion and X-ray is normal, a bone scan should be done as well as a CBC, ESR, SPEP, calcium, phosphorus, alkaline phosphatase, and acid phosphatase.

Most lesions are osteolytic, especially colon and thyroid carcinoma and hypernephroma. Breast, lung, and prostate cancer metastases can be osteoblastic.

The patient should of course be referred to the appropriate specialist for management.

Fractures and Dislocations

Their management is beyond the scope of this book and is best left to the orthopedist. Note should be taken of the fact that fractures of the transverse processes, though often appearing insignificant and treated as such, are associated with a tremendous amount of muscular and ligamentous damage and can result in significant and prolonged pain and disability.

Appendix 11-1   Treatments Used for Low Back Pain

Treatment	When Useful	Comments
Ice	Any	
Heat	Local muscle spasm	Sometimes useful as counterirritant
Aspirin	Any	
NSAI	Any	
Muscle Relaxants	Muscle spasm; or when mild decrease in anxiety helpful	
Steroids (oral)	Acute severe pain; significant or progressive weakness	Taper rapidly
Antidepressants	Chronic	
Bedrest	Acute severe pain; significant or progressive weakness	Otherwise no more helpful than "precautions"
Precautions (avoid bend, twist, lift, prolonged sit, pain producing activity.)	Any	
Ergonomic Teaching	Recurrent or high risk for recurrence	By trained physical therapist best
Vocational Rehabilitation	Chronic	When would be helpful
Weight Loss	Chronic or recurrent, obese	
Exercises-Extension	Acute derangement; dysfunction; precautionary	Be sure doesn't peripheralize
Exercises-Flexion	Recovering from derangement spondylolysthesis	Be sure doesn't peripheralize
Exercises-Abdominal Strengthening	Chronic/Recurrent	
Postural Instruction	Any	Expecially sitting (lumbar roll)
Exercise-Aerobic	Chronic	If not contraindicated
Correct Leg Length Discrepancy	Chronic/Recurrent	Correct one half of measured difference, or enough to level pelvis with patient standing
Physical Therapy Modalities (US, EGS, ETC)	Prolonged	? usefulness
TENS	Chronic	
Mobilization/Manipulation	Prolonged	? usefulness
Traction	Acute severe pain or prolonged	May make better, may make worse

Treatment	When Useful	Comments
Gravity Inversion	Not recommended	Lots of dangers—eye, stroke, etc.
Corset	*Temporary* use with derangement; segmental instability; compression fracture	Prolonged use may lead to dysfunction and weakness
Trigger Point Injection	Chronic, if trigger points found	
Facet Joint Injection	Facet Syndrome	Ideally done under fluoro., but often good results with "neighborhood" injection
Epidural Steroid Injection	Prolonged sciatica or weakness	One of few treatments proved useful by controlled experiment. Should only be done by physician experienced in technique/risks.
Chymopapain	Prolonged sciatica or weakness; CT confirms consistent disc lesion	Done in OR by surgeon
Laminectomy	Same	Not a cure-all; recurrent problems later very common
Foramenotomy	Same, CT shows nerve impingement by hypertrophied facet	
Fusion	Segmental instability	

Your Low Back-Self Care

Here are some things you can do when your back "goes out":

1. Rest. You don't have to stay in bed unless your doctor tells you to, but definitely don't do any bending, lifting or twisting while your back is sore; you may make things even worse. Especially avoid sudden movements.
2. Try to avoid sitting as much as you can, especially for long periods. And if you do sit, be sure you have some support for your low back so that a "hollow" is kept there; that is, don't allow it to bend forward..

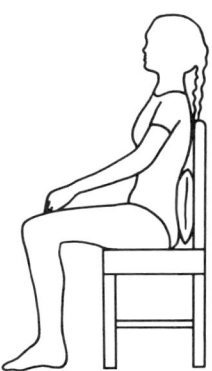

3. Use aspirin or acetaminophen as directed, unless your doctor has prescribed a different medicine.
4. Apply ice to the sore area for twenty minutes or so at a time. Some people find wet heat feels better, and others like to alternate the two.
5. A simple exercise: once an hour put your hands on your hips and arch your low back backwards ten times in a row (don't just bend your knees!). If it hurts to do this, try it lying on your stomach instead of standing. This should

make the pain better; if it makes things worse or sends pain down your leg, stop the exercise and let your doctor know.
6. If something unusual happens or things don't clear up quickly, be sure to contact your doctor.

PATIENT HANDOUT 11-2

Your Low Back: Posture and Precautions

A lot of low back trouble is caused by poor posture, especially when sitting, or by bending or lifting improperly. Here are some tips:

Sitting:

Be sure to sit with a hollow in your low back. For most chairs that means you will have to attach a pillow or cushion to keep you in proper posture. (Don't forget your car!)

Sleeping:

A moderately firm mattress with a very firm box-spring or board under it is best. Most people find it most comfortable to sleep on their side with their legs curled up, or on their back with a pillow under their knees.

Standing:

Avoid positions where you are standing bent over for long periods, even if it's only a little bit bent over. Be especially careful with this when shaving, putting on make-up, doing the dishes, vacuuming, gardening, etc. If you must be bent over for a while, straighten up every few minutes and arch your back a few times.

LIFTING AND CARRYING:

Several rules, *very* important:
1. Never twist to lift something or put it down; turn your feet instead.
2. Bend at the hips and knees, never at the waist.
3. Don't handle things that are too heavy.
4. Hold objects close to your body, not with your arms extended out.
5. Don't lift anything unbalanced.
6. Never carry anything above waist level.

EATING:

Losing some weight might help too!

A SIMPLE EXERCISE:

Put your hands on your hips and arch your low back *backwards* (don't just bend your knees); do ten of these in a row after you have been bent over, or after you have been sitting for a while, or when your back hurts. You can also do this when recumbent, lying face down and pushing up with your arms to arch your back.

If this exercise makes your back hurt more (other than just a "stretching" sensation) or makes pain go down your leg, stop and let your doctor know.

Low Back Exercises

1. **Extension**

   Put your hands on your hips and arch your low back *backwards* (don't just bend your knees); do ten of these in a row after you have been bent over, or after you have been sitting for a while, or when your back hurts. You can also do this when recumbent, lying face down and pushing up with your arms to arch your back.

2. **Flexion**

   Lying down on your back, grab one knee at a time and bring it toward your chest and stretch; alternate knees and repeat ten times for each. As you get more limber, bring both knees up together.

3. **Abdominal Strengthening**

   Strong abdominal muscles are very important for a healthy back.

   Lying down with your hands behind your head, start to do a sit up and tighten your abdominal muscles hard; let go and repeat ten to twenty times. Don't actually sit up very far because that can put a lot of strain on your back.

12

The Coccyx

Evaluation 148
Coccyx Bruise or Fracture 148
Coccydynia 149

EVALUATION

What is the *duration* of the symptoms?
Was there any *trauma*, or was the onset of pain related to childbirth?
Are there any *gastrointestinal* or *gynecologic* symptoms?
Tenderness and *mobility* of the coccyx should be evaluated by performing a digital rectal examination as well as palpating the external aspect.
The *lumbosacral* and *thoracic spine* should be palpated as well.
Be very careful, especially with acute pain, that what you are dealing with is not a *perirectal abscess*, a *pilonidal cyst* or a *neoplasm*.

COCCYX BRUISE OR FRACTURE *(Common)*

Diagnosis

There is usually a history of a fall on the tailbone. The coccyx can sometimes be fractured during childbirth.
Examination reveals coccygeal tenderness, and if a fracture is present, painful excessive mobility of the coccyx on rectal examination.

COCCYDYNIA

If palpation of the more proximal spine is painful, an X-ray of the sore area must be done to rule out an associated compression fracture there.

Treatment

A *ring seat* (a life preserver serves quite well) takes pressure off the injured part.

Analgesics should be used as needed.

The presence of a fracture means that pain will probably persist for months, but no additional specific treatment is indicated or available.

COCCYDYNIA *(Uncommon)*

Diagnosis

That patient gives a history of weeks or months of pain, worse after sitting.

Be sure that there are no symptoms suggestive of gastrointestinal or gynecologic disease.

Examination reveals tenderness to palpation only. (Rectal exam is otherwise normal.)

Rule out the possibility of pain referred from gynecologic or retroperitoneal structures or the lumbosacral spine.

Pathophysiology

This is usually due to previous trauma or to frequent and/or prolonged sitting on hard surfaces. It can also be due to pain referred from the lumbar spine.

Treatment

The patient should *avoid sitting on hard surfaces*.

A *ring seat* is helpful.

Steroid injection may be offered. Infiltrate a cc each of steroid and of local anesthetic around the tender area after sterile prep and ethyl chloride spray. Be sure no infection is present. See p. 22 for further details and precautions.

13

Neurologic Symptoms and Diffuse Pain of the Lower Extremity

Evaluation	150
History	151
Examination	151
Electrodiagnosis	154
X-Ray	155
Syndromes	155
Pseudoclaudication	155
Meralgia Paresthetica	156
Common Peroneal Nerve Compression	157
Deep Peroneal Nerve Compression	157
Tarsal Tunnel Syndrome	157

EVALUATION

Neurologic symptoms and especially pain of the lower extremity can be a manifestation of *vascular disease*. The diagnosis and management of this problem is beyond the scope of this book, but its possibility must be kept in mind and appropriate history taken and examination and tests performed to exclude it.

Also, weakness, numbness, and paresthesias obviously are often due to lesions of the *brain* or *spinal cord*. A *polyneuropathy* will usually manifest itself as sensory loss in a "stocking" distribution without associated pain, but metabolic disease (especially diabetes) can affect single large nerves and lead to confusion with the compression neuropathies discussed herein. And diseases of *muscle* or the *neuromuscular junction* can underlie motor weakness. Again, discussion of these conditions is well beyond the scope of this volume; but it is essential that whatever history, examination and tests are necessary to rule out such an etiology be performed whenever the initial history or examination are at all suggestive of such a possibility.

Occasionally neurologic symptoms and diffuse pain of the lower extremity can be due to nerve root impingement by *pelvic* or *retroperitoneal pathology*, a point that should be kept in mind.

History

Ask about the *duration* of symptoms and the *location* of *pain* and *paresthesias*. Often the patient's response to this question is too vague to be useful. See Figure 13-1 (pp. 152-153).

Has the patient noticed any *weakness* or *gait difficulty*?

Inquire about *precipitating factors*. Aching pain which comes on with a fairly consistent amount of walking and is then relieved by rest is of course suggestive of *intermittent claudication* due to vascular compromise. If vascular examination is normal even after exercise, the patient may have the PSEUDOCLAUDICATION SYNDROME, discussed below. A similar pattern of exacerbation with activity may be seen with pain referred from nerve root irritation in the back or an intermittent COMPARTMENT SYNDROME (discussed in Chapter 21).

Find out about any *back* or *buttock pain* and its association with the more distal symptoms.

Be sure to ask about any *weakness, numbness,* or *paresthesias* elsewhere in the body, or any headaches, visual or auditory symptoms, vertigo, imbalance, incoordination, speech difficulty, episodes of loss of consciousness, or recent head trauma. Any of these may of course suggest a central nervous system lesion.

Examination

Examine the *back* and do the *straight leg raising* test and other tests as necessary to see if the symptoms are reproduced or exacerbated. See Chapter 11 for a description of these maneuvers. Often distal pain and/or paresthesias are the only manifestations of nerve root compromise in the low back, back pain being conspicuous by its absence. As a matter of fact *this etiology is much more common than any of the other syndromes discussed in this chapter*. If such is the case, management is as discussed in Chapter 11.

Examine the *hip, knee,* and *ankle* for range of motion, swelling, and tender-

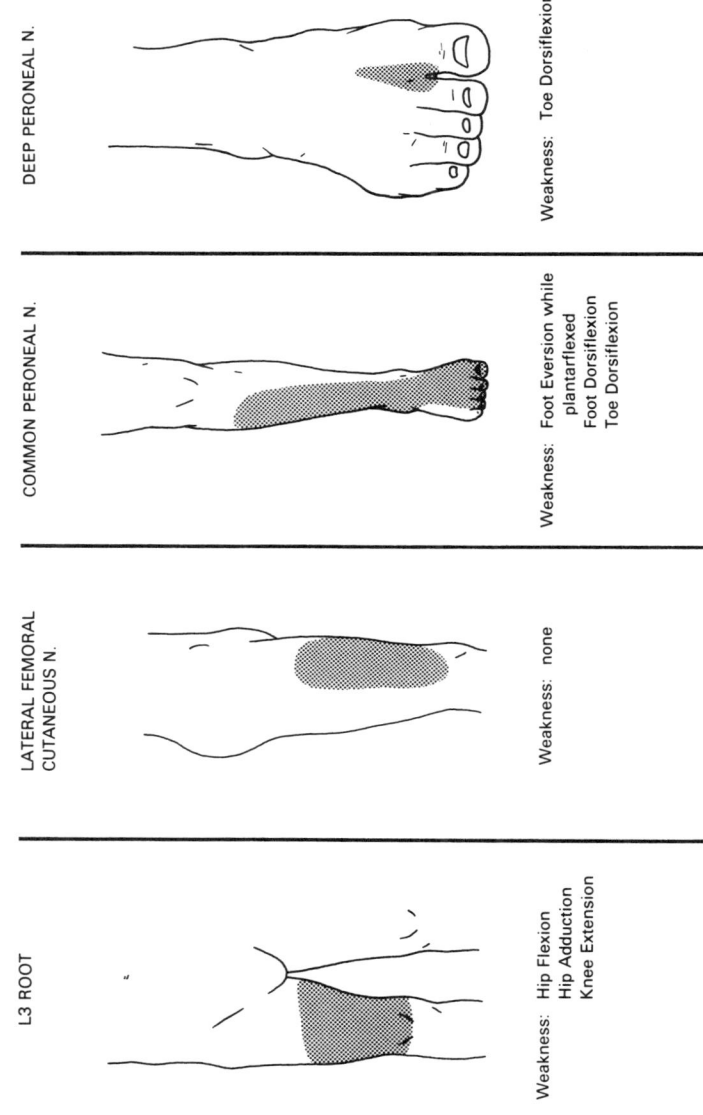

Fig. 13-1. Common root and peripheral nerve syndromes (lower extremity). Syndromes are usually incomplete. Dermatomes and innervations can vary in some patients.

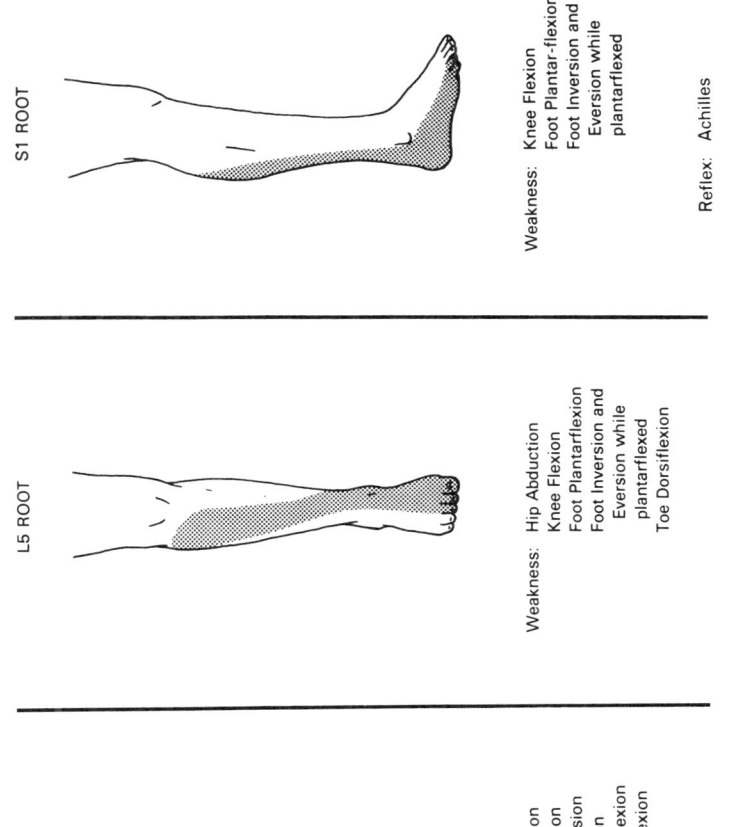

L4 ROOT

Weakness: Hip Adduction
Hip Abduction
Knee Extension
Knee Flexion
Foot Dorsiflexion
Toe Dorsiflexion

Reflex: Patellar

L5 ROOT

Weakness: Hip Abduction
Knee Flexion
Foot Plantarflexion
Foot Inversion and Eversion while plantarflexed
Toe Dorsiflexion

S1 ROOT

Weakness: Knee Flexion
Foot Plantar-flexion
Foot Inversion and Eversion while plantarflexed

Reflex: Achilles

154   13. NEUROLOGIC SYMPTOMS AND DIFFUSE PAIN OF THE LOWER EXTREMITY

ness, since often a specific articular or periarticular condition will be the cause of the patient's complaints even though he or she is unable to localize the symptoms.

Look for *tenderness* and *tenseness* in the lower leg, and if present see if the pain is exacerbated by passive stretch of the involved compartment. If so the patient may have a COMPARTMENT SYNDROME which if acute is a true orthopedic emergency, requiring STAT referral.

Check distal *pulses* and *capillary filling*.

Elicit and knee and ankle *reflexes* (Figure 13-2).

Examine *strength*. See Table 13-1 for various motions to be tested and their root and peripheral innervations. *Significant* or *progressive weakness* is always an indication for referral for more precise diagnosis (e.g., EMG) and more aggressive management, and may be suggestive of a CNS lesion.

Look for muscle *atrophy*.

Test for *pin sensation*. When a more sensitive examination needs to be done, check *two-point discrimination*. See Figure 13-1 for the areas classically involved with the commonly impaired lumbosacral roots and peripheral nerves; but note that areas are often incompletely involved, that sensation testing is subjective and thus often quite inaccurate unless the deficit is almost complete, and that there are often individual variations from the dermatomes shown.

Electrodiagnosis

When the diagnosis cannot be clearly defined by the history and examination described above, referral should be made for EMG-NCV. See Chapter 2.

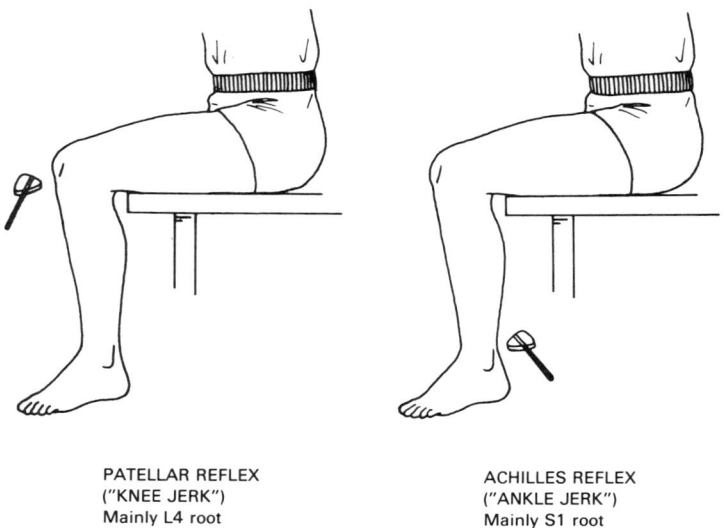

PATELLAR REFLEX
("KNEE JERK")
Mainly L4 root

ACHILLES REFLEX
("ANKLE JERK")
Mainly S1 root

Fig. 13-2.   Reflexes: lower extremity.

SYNDROMES

Table 13-1   Strength Testing: Lower Extremity

All maneuvers are done against your resistance, and with the patient seated with legs dangling. Compare with the uninvolved side.

- FLEXING THE HIP (raising the thigh)
  (ILIOPSOAS; L1-2-3; femoral n.)
- ADDUCTING THE HIPS (bringing the knees together)
  (ADDUCTORS; L3-4; obturator n.)
- ABDUCTING THE HIPS (spreading the knees apart)
  (GLUTEI; L4-5-S1; superior gluteal n.)
- EXTENDING THE KNEE (Straightening the leg)
  (QUADRICEPS; L2-3-4; femoral n.)
- FLEXING THE KNEE (pulling heel under table)
  (HAMSTRINGS; L4-5-S1; sciatic n.)
- PLANTAR FLEXING THE FOOT (in neutral)
  (GASTROCNEMIUS; L5-S1-2; tibial n.)
- HOLDING THE PLANTAR FLEXED FOOT INVERTED
  (POSTERIOR TIBIALIS; L5-S1; tibial n.)
- HOLDING THE PLANTAR FLEXED FOOT EVERTED
  (PERONEI; L5-S1; sup. peroneal n.)
- DORSI FLEXING THE FOOT (in inversion)
  (ANTERIOR TIBIALIS; L4; deep peroneal n.)
- DORSI FLEXING THE GREAT TOE
  (EXTENSOR HALLUCIS LONGUS; L4-5; deep peroneal n.)

X-Ray

A lumbosacral spine X-ray should be obtained in every patient with objective neurologic findings not clearly due to a peripheral problem. CT scanning and referral is warranted in cases with severe or progressive deficit or if there is suspicion of a spinal cord lesion.

SYNDROMES

Pseudoclaudication (spinal claudication, spinal stenosis)
(Uncommon)

Diagnosis

    The patient complains of leg pain, usually bilateral and sometimes associated with neurologic symptoms, which comes on with walking.

    Usually in older patients.

The patient usually reports that sitting down relieves the symptoms as opposed to just stopping giving relief in patients with vascular claudication; and symptoms are usually slower to come on and slower to resolve.

On examination, normal pulses and vascular status are found in the lower extremities, even after exercise and during symptoms.

The stationary bicycle test is diagnostic; patients with spinal claudication can usually pedal all day as long as the back remains flexed, while those with arterial insufficiency won't have exercise tolerance any greater than with walking.

(The picture may be greatly complicated by the coexistence of vascular insufficiency and pseudoclaudication).

CT scan of the lumbar spine is necessary to measure the diameter of the spinal canal and to rule out other lesions.

Pathophysiology

Narrowing of the spinal canal occurs because of degenerative bony hypertrophy and protruding disc material, usually superimposed on a congenitally narrow canal.

Prolonged extension of the back, as in walking, leads to compression of the cauda equina or its blood supply and thus the symptoms.

Treatment

Other than *symptomatic* treatment, only *surgical decompression* of the spinal canal can help. This should be resorted to only if symptoms warrant.

Meralgia Paresthetica *(Uncommon)*

Diagnosis

Numbness and often a burning sensation is felt in the lateral aspect of the thigh (Figure 13-1).

It must be differentiated from L4 and L3 root lesions.

If there is no evident cause, a diabetic mononeuropathy must be ruled out by checking blood sugars.

Pressure over the anterior iliac spines may cause tingling in the involved area of the thigh.

Pathophysiology

Compression of the lateral femoral cutaneous nerve occurs as it passes under the inguinal ligament just inferior to the anterior superior iliac spine.

This usually occurs in obese patients or those who wear constrictive girdles or heavy tool belts.

SYNDROMES

Treatment

Avoid compression by the responsible garment or belt by discontinuing its use, moving it or padding it. Advise *weight loss*.

Common Peroneal Nerve Compression *(Rare)*

Diagnosis

Foot drop is the most dramatic result. See Figure 13-1.
Tenderness is found at the fibular head.

Pathophysiology

Compression usually occurs where the nerve passes over the fibular head, either a result of trauma or of sitting "indian style" with the outsides of the knees on a hard floor.

Treatment

The nerve must be decompressed by treating the underlying cause; *referral* for this (as well as for bracing and rehabilitation for any foot drop) should be made.

Deep Peroneal Nerve Compression *(Uncommon)*

Diagnosis

Decreased sensation is found between the first and second toes, and sometimes mild weakness of the extensor of the great toe as well (Figure 13-1).
Tenderness occurs where the nerve becomes superficial in the anterior ankle.
This must be differentiated from common peroneal nerve injury and from the anterior compartment syndrome.

Pathophysiology

Compression at the anterior ankle occurs by shoes that are too tight or too tightly laced.

Treatment

Simply by *avoidance* of that compression.

Tarsal Tunnel Syndrome *(Rare)*

Diagnosis

A burning pain is felt in the sole of the foot, sometimes with radiation up the back of the leg.

Sensory loss can be found on the inferior heel or the sole of the foot. The specific area is variable depending on which branch of the nerve is most affected.

Tenderness is found over the nerve just behind the medial malleolus.

Pathophysiology

Compression of the posterior tibial nerve occurs as it passes under the lancinate ligament just posterior to the medial malleolus. The cause may be trauma, direct compression by a shoe, or excessive pronation of the foot.

Treatment

Correct the *underlying cause* as much as possible (especially excessive pronation).

Injection may be tried; technique is to prep, spray with ethyl chloride, and then to infiltrate ½ cc each of steroid and local anesthetic without epinephrine just behind the medical malleolus. See Chapter 3 for further details and precautions.

If these measures fail or if weakness of the toe flexors at the proximal joints develops, the patient should be referred for electrodiagnostic confirmation and *surgical release*.

14

The Hip and Thigh

Evaluation	159
History	159
Examination	161
X-Ray	163
Hip Pain and Strain	163
Trochanteric Bursitis	163
Ischeal Bursitis	164
Groin Pull	165
Hip Pointer	165
Hamstring Pull	166
Degenerative Arthritis of the Hip	166

Hip pain in *children* is covered in Chapter 19. Hip pain in *runners* is covered in Chapter 20. Further details on therapeutic suggestions made in this chapter are to be found in Chapter 3.

For the anatomy of the hip see Figure 14-1.

EVALUATION

Always consider the possibility of hip pain being *referred* from pelvic, intra-abdominal or retroperitoneal disease. Although the differential diagnosis of these conditions is obviously beyond the scope of this book, they must be kept in mind and appropriate history, examination and tests done to rule them out, especially if the initial history and/or examination is not completely consistent with a mus-

14. THE HIP AND THIGH

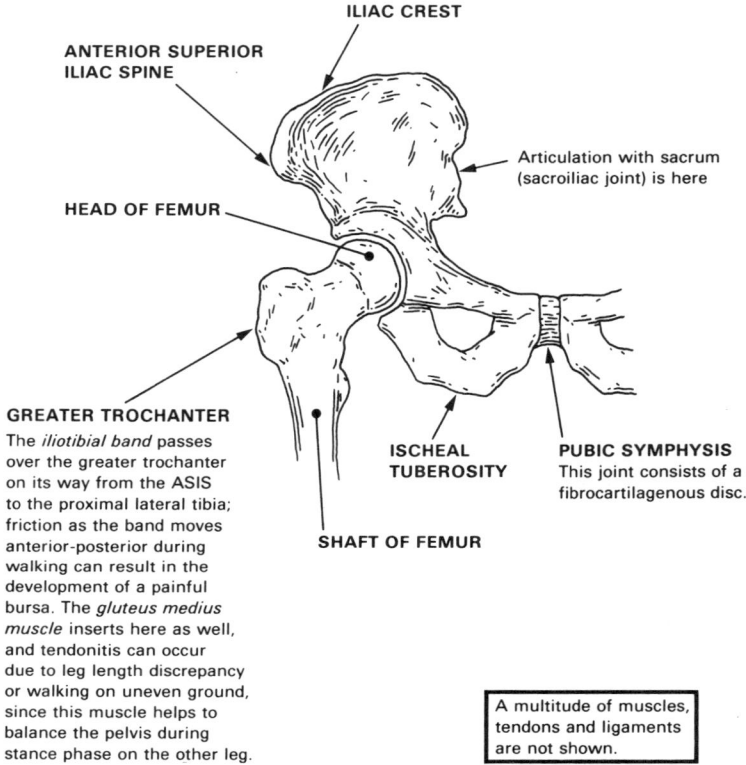

Fig. 14-1. Anatomy of the hip.

culoskeletal origin of the patient's pain. And remember that *deep vein thrombosis* can occur in the thigh and groin.

Aseptic necrosis is rare in adults (except following fracture or in patients with sickle cell disease) but can occur.

History

Ask about *duration* and *location* of pain. Pain from within the hip joint is usually felt either in the groin or the buttock, though it can be referred to the thigh or the knee.

What are *precipitating* and *relieving factors*?

Has there been any specific *trauma, strain,* or *overuse*?

Is there any *back pain* or *trauma,* any *radiation* of pain into the leg, or any *neurologic symptoms* in the thigh or leg? (If so, see Chapter 11 on the low back, or Chapter 13 on neurologic symptoms and diffuse pain of the lower extremity, as appropriate.)

EVALUATION

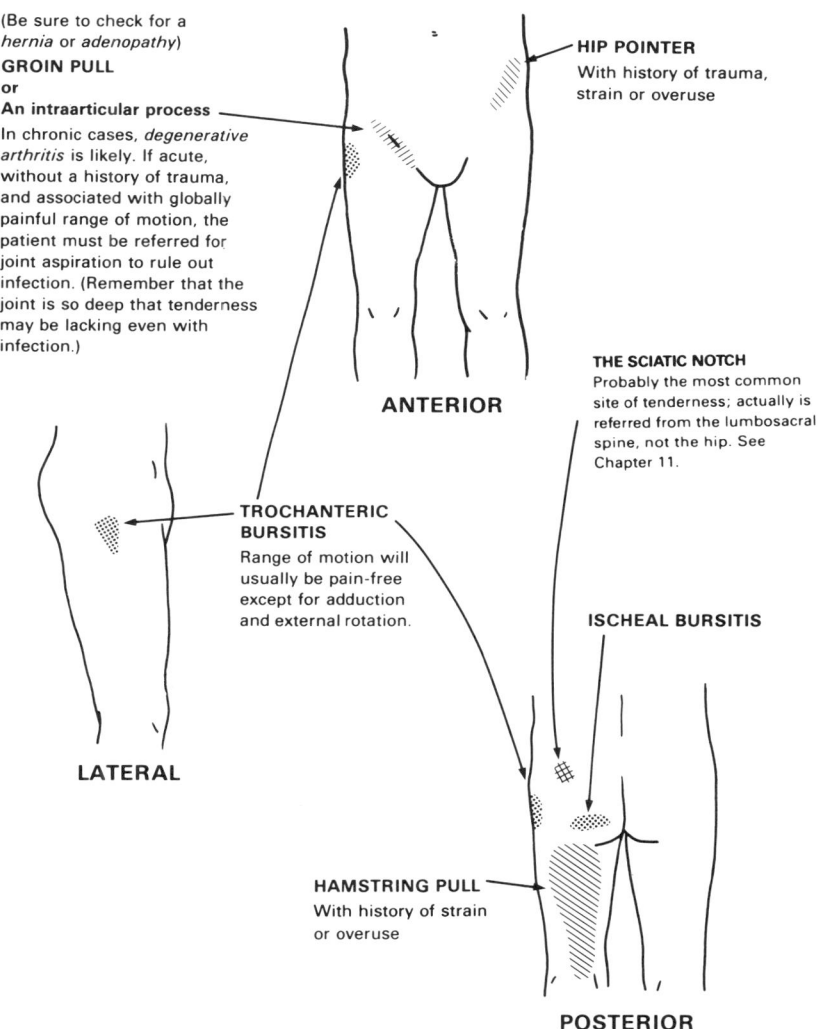

Fig. 14-2. Sites of tenderness around the hip.

Inquire about *previous* hip problems, diagnoses, or treatments.

Are any *other joints* involved? Is there a past history of arthritis or gout?

A complaint of "snapping" in the hip is most commonly due to the iliotibial band popping over the greater trochanter. (If so, the snapping will be palpable there as the hip is flexed and extended.)

Examination

Localize *tenderness* (Figure 14-2).
Observe the patient's *gait*.

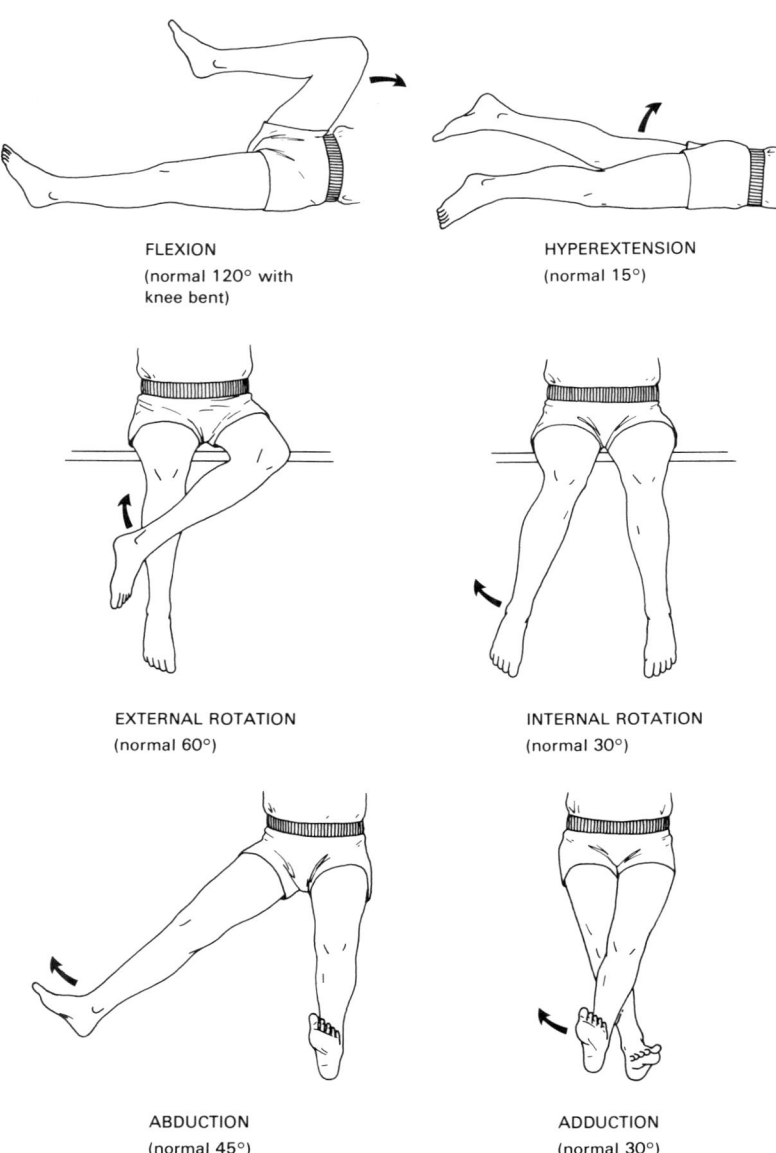

Fig. 14-3. Hip range of motion.

HIP PAIN AND STRAIN

Check *range of motion* (Figure 14-3). Pain in all directions of motion implies an intraarticular process, possibly a septic joint or a fracture in acute cases.

Checking *resisted* range of motion in the same directions will of course indicate strength of the muscles about the hip. In musculotendonous strains, passive stretch and resisted contraction of only the involved muscles will be painful.

Examine the *lumbosacral spine* and do the *straight leg raising test* to help rule out referred pain from the low back. Also examine distal *neurovascular* status. The most common cause of the complaint of hip pain in the postadolescent presenescent patient is lumbosacral disease. See Chapter 11 if appropriate.

Measure *leg lengths* (from each anterior superior iliac spine to its medial malleolus) and also distances between the ASIS and the greater trochanters to detect hip joint shortening.

Do *abdominal, pelvic,* and *rectal* examination as appropriate.

X-Ray

Although when X-rays should be taken was discussed in Chapter 2, the criteria must be looser in the case of the hip because it is so inaccessible to palpation and is so often the site of metastatic disease. Any painful hip in which the diagnosis is not obviously both periarticular (e.g., trochanteric bursitis) and rapidly resolving should be X-rayed.

Note that femoral neck fractures can be difficult to pick up on early X-rays, and the consequences of missing one can be disastrous. Any patient with hip trauma and significant pain, limitation of motion or unwillingness to bear weight, should be considered to possibly have a fracture even in the presence of apparently normal x-rays. Orthopedic consultation should be obtained.

HIP PAIN AND STRAIN

Trochanteric Bursitis *(Common)*

Diagnosis

Tenderness is found over the greater trochanter only (Figure 14-2).
Pain is elicited with external rotation and adduction only.
It is very commonly found in association with sciatica.

Pathophysiology

Several muscles attach at the greater trochanter; stresses at these insertions can lead to inflammation.

Or, a painful bursa can develop due to friction between the trochanter and the overlying iliotibial band.

Treatment

Try *rest, ice* or *heat*, and *antiinflammatory medications*.

Steroid injection has a very good success rate and should be offered if needed. (See Figure 14-4.)

Referral to a physical therapist for instruction in iliotibial band and hip rotator stretching as well as antiinflammatory modalities may be helpful.

Ischeal Bursitis (Weaver's Bottom) *(Uncommon)*

Diagnosis

The patient gives a history of prolonged sitting, especially with the legs crossed and on a hard surface. Common in cyclists.

Tenderness is found only over the ischeal tuberosity (Figure 14-2).

Pathophysiology

Inflammation is due to prolonged pressure.

Treatment

Have the patient *avoid prolonged sitting*, especially with the legs crossed.
The use of a *soft pad* or a *ring seat* may be helpful.
Antiinflammatory medications can be used if the problem is acute.

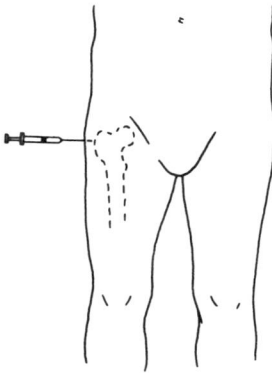

Fig. 14-4. Injection for trochanteric bursitis. Prep over the tender area and spray with ethyl chloride. use a 23-gauge needle to inject a cc each of steroid and local anesthetic in several directions without withdrawing from the skin) around the inflamed area. See pp. 22-24 for further details and precautions.

HIP PAIN AND STRAIN

Groin Pull *(Common)*

Diagnosis

The patient reports acute or repetitive abduction strain.
Tenderness may be present over the groin or the medial thigh.
Pain will be found on passive stretch or resisted contraction of only the involved muscles.
Be sure to check that the patient does not have a hernia or lymphadenitis and that findings are not suggestive of an intraarticular process.

Pathophysiology

Strain is of the hip flexor and/or the adductor musculature.

Treatment

Use *ice* if acute.
Advise *rest*.
Antiinflammatory medications may hasten resolution if the pull is at all severe.
As the episode is resolving, stretching and strengthening exercises are imperative to prevent recurrence. Consider physical therapy referral, especially in athletes.

Hip Pointer *(Common only in athletes)*

Diagnosis

There often is history of a direct blow to the iliac crest region. Alternatively, the patient may have been doing much running.
An X-ray must be done to rule out an avulsion fracture of the anterior superior iliac spine or iliac crest.

Pathophysiology

A painful hematoma develops from direct trauma in the soft tissues that attach there.
Or this can be a traction injury to muscular origins.

Treatment

Use *ice* and *compression* acutely.
In teenagers, avoidance of strain and trauma is advisable for six to eight weeks. Older patients may return to activity as tolerated.

Hamstring Pull *(Common)*

Diagnosis

The patient gives a history of acute or repetitive strain.
Pain and tenderness may be located anywhere in the posterior muscles of the thigh.

Pathophysiology

Often the basic problem is quadriceps muscles relatively too strong for the opposing hamstring muscles.

Treatment

Use *ice* frequently when acute, as needed later.
Rest is advisable.
Apply a compressive bandage (crutches if severe), with resumption of activity as tolerated.
Antiinflammatory medications may hasten resolution if the pull is at all severe.
Once the episode has resolved, *strengthening* and *stretching exercises* to prevent recurrence are advised. Physical therapy referral may be best.

Degenerative Arthritis (Osteoarthritis) of the Hip *(Very Common)*

Diagnosis

Chronic pain is worse after weight-bearing activity, and is better with rest. The pain can be felt in the groin, the buttock, the thigh, or the knee.
Found in older patients or those with previous significant hip disease or trauma.
There may be variable pain with range of motion and with gait.

Pathophysiology

See Chapter 22.

Treatment

A proper balance of rest and use must be found.
Antiinflammatory medications may be used.
In cases with significant symptoms, refer to a physical therapist for a range of motion and muscle strengthening program.
Weight loss is critical if the patient is obese.
The use of a *cane* can be quite helpful for some patients.
Referral for consideration of *injection* or *surgery* should be considered in severe cases not responding to conservative management.

15

The Knee

History 168
Examination 171
Knee Pain 182
    Evaluation 182
    Patellofemoral Pain 183
    Recurrent Patellar Subluxation 186
    Plica Syndrome 187
    Breaststrokers Knee 187
    Meniscus Tears 188
    Patellar Tendonitis 189
    Anserine Tendonitis 190
    Degenerative Arthritis of the Knee 190
Knee Giving Way and Catching/Locking 191
    Evaluation 191
    Chronic Ligamentous Instability 192
Knee Swelling 193
    Evaluation 193
    Prepatellar Bursitis 193
    Infrapatellar Bursitis 194
    Baker's Cyst 194
Knee Trauma 195
    Evaluation 195
    Collateral Ligament Sprain 195
    Anterior Cruciate Ligament Injury 196
    Posterior Cruciate Ligament Injury 197
    Acute Patellar Dislocation 197
    Patellar Contusion 197

Knee problems in *children* and *adolescents* are covered in Chapter 19. Knee pain in *runners* is discussed in Chapter 20. Further details on therapeutic suggestions made in this chapter can be found in Chapter 3.

For the anatomy of the knee see Figure 15-1.

HISTORY

Ask about the *duration* and *location* of pain.

What are *precipitating* and *relieving factors*?

Has there been any recent or previous *injury*? If so, what was the mechanism thereof?

If there was swelling after the injury, ask whether it occurred immediately (a sign of hemarthrosis, which in turn usually means a fracture or cruciate ligament tear), or a few hours later (which would be suggestive of an intraarticular lesion irritating the synovium, such as a meniscus tear).

If there was a noise at the time of injury, a pop is said to indicate an anterior cruciate ligament tear, a crunch to indicate a meniscal tear, and a tearing sensation to be suggestive of a collateral ligament injury. These clues are not completely reliable.

Ask about any recurrent *swelling,* and whether it is associated with *redness* or *warmth*.

Does the knee ever *lock*? A locked knee is one that cannot be fully extended because of a physical block, not just because of pain or swelling; it is usually due to a torn meniscus. A patient seen after trauma with a locked knee should be seen by an orthopedist within 24 hours. Recurrent *catching* or *clicking* of the knee may be due to a loose body or torn meniscus, or sometimes to degenerative arthritis. *Snapping* often is caused by an inflamed synovial plica.

Does the knee *give way*? Occasional giving way or almost giving way is common with patellofemoral pain or symptomatic degenerative arthritis; true giving way, especially if followed by a period of pain and swelling, usually indicates a subluxing patella, a loose body due to previous trauma or osteochondritis dissecans, a torn meniscus, or (especially if it only occurs when stepping a certain way) to chronic ligamentous instability.

Is there a *grating* sensation? (Common with patellofemoral pain or degenerative arthritis).

Inquire about any *previous* knee problems, diagnoses, or surgery.

Find out how the problem *limits* the patient in his or her daily activities.

Are there *other joints* involved, or is there a history of arthritis or gout?

Always remember that knee pain may be *referred* from the hip or back. As a matter of fact, *the most common significant cause of knee pain in a child is pain referred from serious disease in the hip;* and a similar situation may occur in *older patients*.

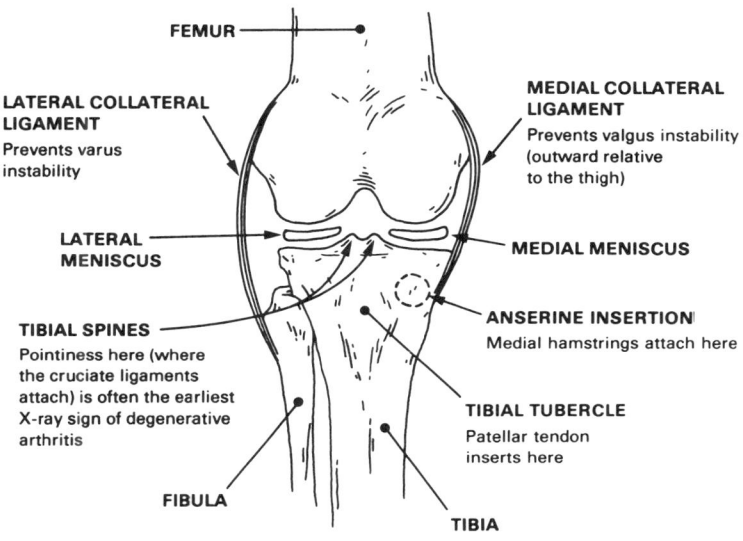

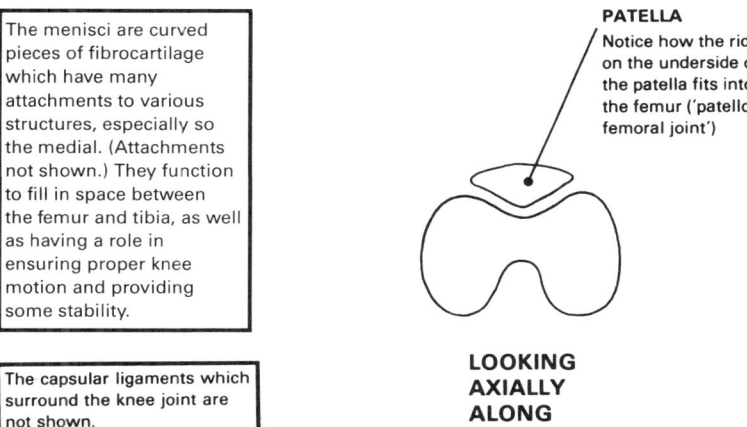

Fig. 15-1. Anatomy of the knee.

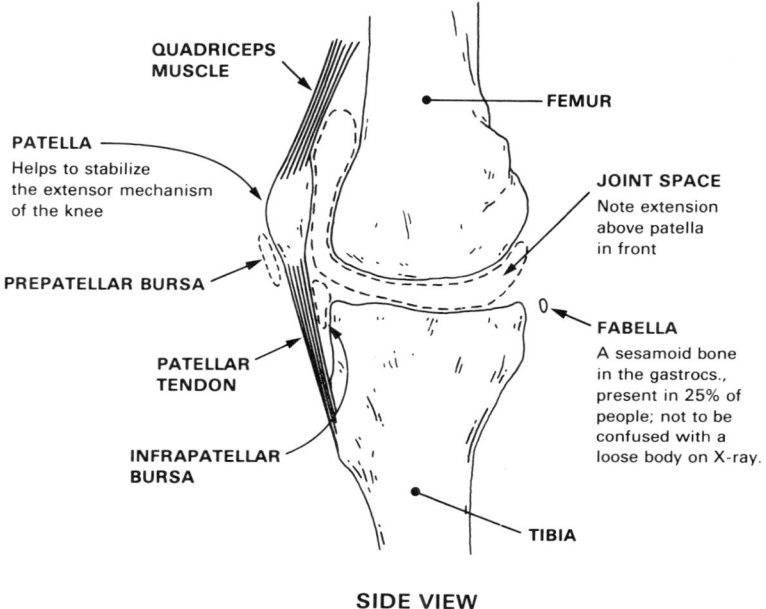
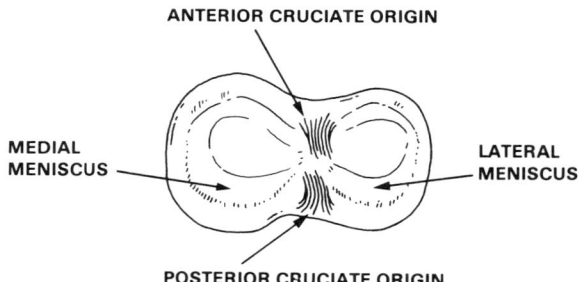

Fig. 15-1. Continued.

EXAMINATION

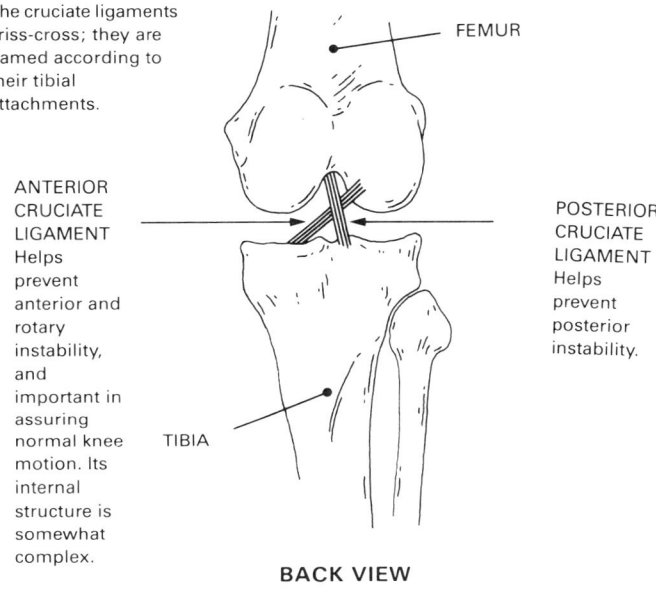

Fig. 15-1. Continued.

EXAMINATION

A complete knee examination may not be required for every patient with knee complaints. As experience is gained, it will become more obvious earlier in the evaluation of the patient what the diagnostic possibilities are, and attention can be concentrated on those findings relevant to the particular cause at hand.

Unless there is acute pain or recent significant injury, evaluating the *functional status* of the knee is a good first step; it can often give a good idea of the severity of the problem and set the tone for the rest of the examination. This is done by observing the patients *gait*, by asking the patient to *hop* on first the unaffected and then the affected leg, and by having him or her do a *deep knee bend* on each leg while holding on for balance. If unsure as to whether these maneuvers are safe, this portion of the examination should be deferred to the end.

Observe the *alignment* of the limb, i.e., how much valgus or varus deformity exists. This is especially significant to note in patients with degenerative arthritis.

Assessing *muscle tone* is next. The *thigh circumference* should be measured approximately six centimeters proximal to the superior patellar pole on both legs. More than a centimeter of difference is significant. Atrophy can develop fairly quickly and indicates that there is indeed a significant problem with the involved leg. Observing the *tone* of the contracted quadriceps can give an indication of whether muscle rehabilitation is indicated; this is important in the treatment of almost all knee symptoms, and in the prevention of injury as well. Look specifically at the VMO bulge; inadequate tone here is an important contributing factor to anterior knee pain. (See Figure 15-5).

Look for *effusion,* which will first be noticeable as a loss of the groove at the

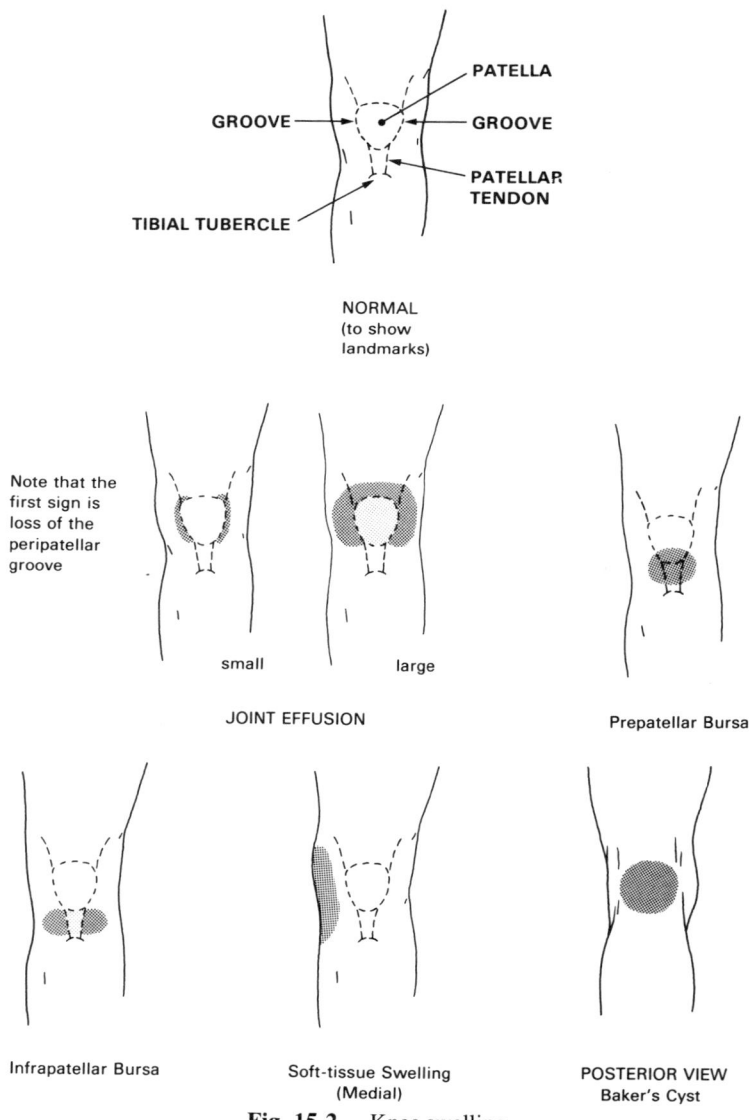

Fig. 15-2. Knee swelling.

EXAMINATION

medial side of the patella. Joint effusion must be distinguished from swelling in the prepatellar or infrapatellar bursae, from a Baker's cyst, and from soft tissue swelling localized only to the traumatized area of the knee (Figure 15-2). An atraumatic effusion may mean infection, gout, or pseudogout and must be tapped (Figure 15-3). Remember that the presence of *redness* or *warmth* may mean infection.

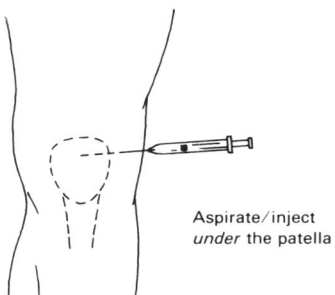

Fig. 15-3. How to aspirate or inject a knee joint. Use sterile technique throughout. After proper preparation of the site, raise a wheal of local anesthesia under the superior-lateral aspect of the patella. Then inject some of the anesthetic in the direction shown. Now withdraw and use an 18-gauge needle on a syringe of at least 20 cc capacity (a smaller syringe does not create enough vacuum on aspiration) in the direction shown. If steroid is to be injected, use a hemostat to hold the needle bevel and change syringes to one containing 2 to 6 cc of steroid. (Never inject steroid if the synovial fluid is cloudy or if there is other reason to suspect infection.) Steroid can be injected in the same way, without aspiration, when indicated, using a 22-gauge needle. Be sure to see pp. 13 and 24 for discussion of synovial fluid analysis and of precautions and risks of steroid instillation.

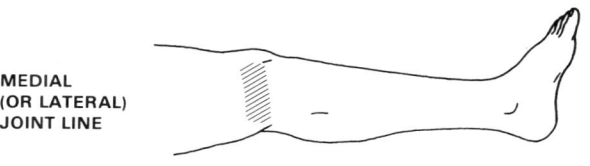

MEDIAL
(OR LATERAL)
JOINT LINE

The groove between the femur and the tibia is palpable.

Tenderness here can be in the ligament, in the meniscus or can indicate degenerative arthritis.

The point of maximal tenderness will usually move posteriorly a bit when the knee is flexed if a painful *MENISCUS* is responsible (called Steinman's test.)

Tenderness may extend a bit above and below the joint line if it arises from the *LIGAMENT*.

Joint-line tenderness due to *DEGENERATIVE ARTHRITIS* is usually fairly diffuse.

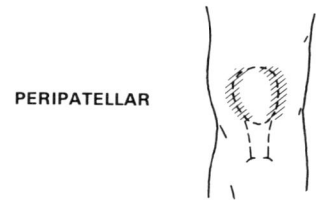

PERIPATELLAR

Tenderness here can be found in *CHONDROMALACIA PATELLAE* or *DEGENERATIVE ARTHRITIS* of the patello-femoral joint, but the patellar compression test is more specific.

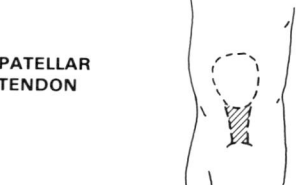

PATELLAR
TENDON

Tenderness anywhere here indicates *PATELLAR TENDONITIS* (except at either end in a pre-teenager or adolescent: See *OSGOOD-SCHLATTER'S DISEASE* in Chapter 19).

Fig. 15-4. Sites of tenderness around the knee.

EXAMINATION

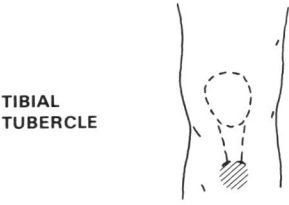

TIBIAL TUBERCLE

Tenderness (and sometimes swelling) here in a teenager implies *OSGOOD-SCHLATTER'S DISEASE* (covered in Chapter 19).

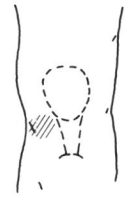

The bony prominence where the anserine tendons insert is easily palpable inferior to the joint line on the antero-medial tibia.

Tenderness here indicates *ANSERINE TENDONITIS*.

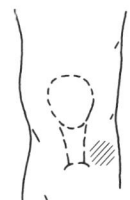

GERDY'S TUBERCLE on the anterolateral tibia

Tenderness here indicates *ILIOTIBIAL BAND TENDONITIS* (covered in Chapter 20).

Fig. 15-4. Continued.

Test *range of motion*. Normal full flexion is 130°. Limitation of motion due to muscle spasm or inadequate relaxation must be distinguished from that due to effusion or to true physical limitation (as by a torn meniscus, for example). Note that a very swollen knee will be held at about 30° of flexion, since in this position the volume of the synovial space is greatest. A knee that can't be actively extended as far as passively is said to have an *extensor lag,* a nonspecific but useful sign of dysfunction.

Look for locations of *tenderness*. See Figure 15-4.

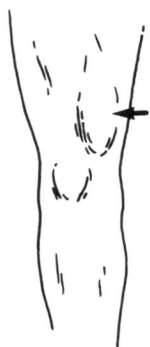

Fig. 15-5. Vastus medialis bulge.

Check *patello-femoral function* by performing the patellar compression and/or patellar inhibition tests (Figure 15-6). Positive results imply the presence of PATELLOFEMORAL PAIN in younger patients, DEGENERATIVE ARTHRITIS of the patello-femoral joint in older ones.

A positive patellar apprehension test (Figure 15-6) or excessive lateral mobility of the patella is suggestive of RECURRENT PATELLAR SUBLUXATION if the history is compatible.

Check the *stability* of the *collateral ligaments* by stabilizing the femur with the knee flexed at 30° (because in full extension other structures are tight), and then exerting a valgus (to test the medial collateral) and then a varus (lateral collateral) stress (Figure 15-7). Getting complete muscular relaxation is the trick. Always compare to the other knee. In acute sprains where you cannot be completely certain that there is no laxity, the knee should be immobilized in a posterior splint and seen by an orthopedist within 72 hours.

Check the integrity of the *cruciate ligaments* by performing the drawer tests (Figure 15-8). In acute injuries, performing the anterior drawer test at 20 degrees of knee flexion (called Lachman's test: see Figure 15-10) has been shown to be much more accurate than at 90 degrees. In chronic instability cases, perform the anterior drawer test with the foot in external rotation as well as in internal rotation and neutral. The *pivot shift test* (Figure 15-9) may also be used; it is somewhat more sensitive than the anterior drawer test for detecting chronic anterior cruciate laxity.

Test for meniscal tears by performing *McMurray's test* (Figure 15-11), which however is often negative even with a tear. If the patient can do a good *deep knee bend* or *duck-walk* (Figure 15-12) without difficulty, medial meniscal problems are unlikely.

Have the patient stand or roll over prone and examine the popliteal area.

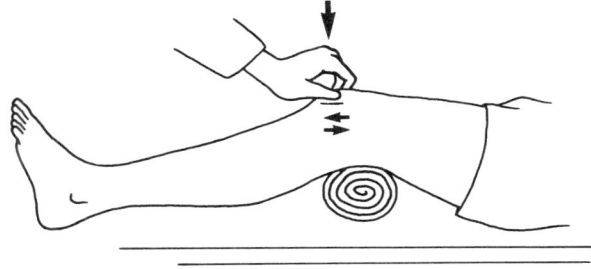

Patellar Compression
Press the patella down into the femur and move it proximally and distally. Pain and/or crepitus is a positive test.

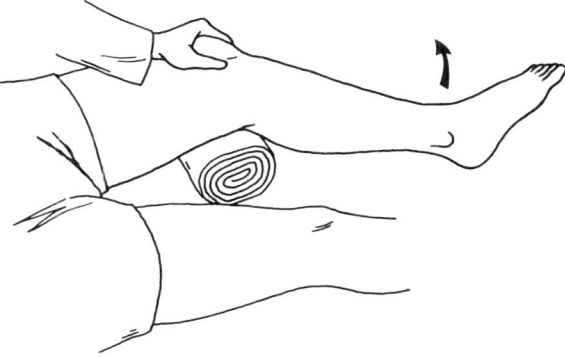

Patellar Inhibition
With the knee flexed about 20° (over your other hand or a rolled towel), put your thumb and forefinger around the top of the patella and ask the patient to straighten his leg. If he starts to, and then stops because of pain, the test is positive.

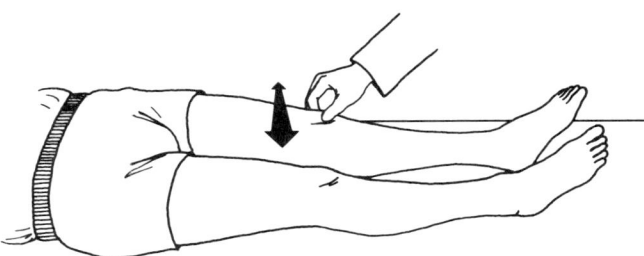

Patellar Apprehension
Have the patient relax and then start to move his patella laterally. If he resists, the test is positive.

Fig. 15-6.   Patello-femoral tests.

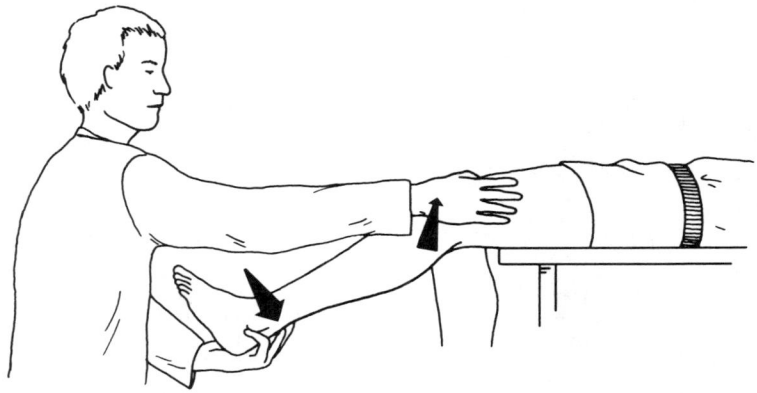

MEDIAL COLLATERAL LIGAMENT

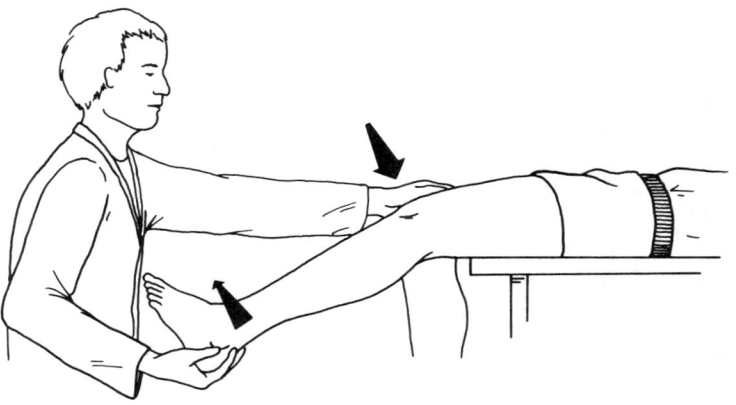

LATERAL COLLATERAL LIGAMENT

Good muscular relaxation is essential
Don't allow the thigh or leg to rotate while performing the test
Always compare to the uninvolved knee

Fig. 15-7. Checking collateral ligament stability.

In patients in whom patellofemoral pain or recurrent patellar subluxation is suspected, it is also helpful to assess *patellar tracking*. This is done by observing the motion of the patella as the knee is flexed and extended, and by having the patient contract his quadriceps while the knee is extended, again watching the patella. Motion laterally (the ''J sign'' or ''comma sign'') or medially and then laterally (the ''C sign'') indicates tracking difficulties. Also important to assess

| ANTERIOR DRAWER TEST | POSTERIOR DRAWER TEST |
| (for the ANTERIOR CRUCIATE LIGAMENT). | (for the POSTERIOR CRUCIATE LIGAMENT). |

Fig. 15-8. The drawer tests. Getting complete muscular relaxation is necessary for a good exam. Always compare to the uninvolved leg. Asymmetric laxity implies a torn cruciate ligament. When doing the anterior drawer test in cases of chronic instability, only one side of the tibia may "rotate" forward rather than the entire tibia translating anteriorly; this indicates "rotatory instability," classified as anterolateral (lateral tibia comes forward) or anteromedial (medial tibia moves forward).

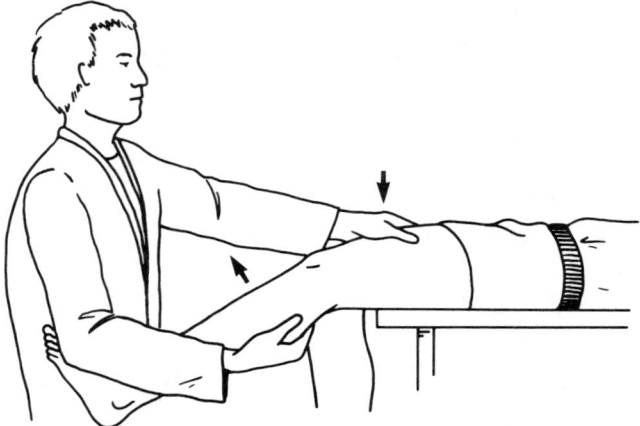

Fig. 15-9. Lachman's Test. Always be sure to obtain good muscular relaxation and to compare to the uninvolved side for an accurate result.

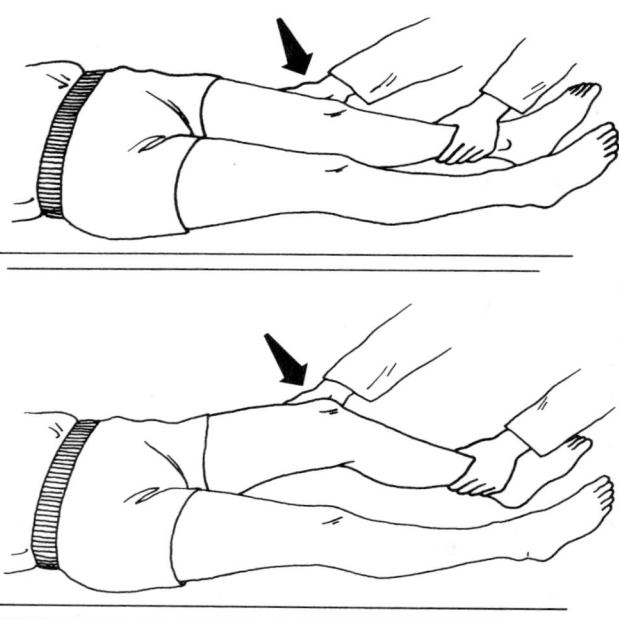

Fig. 15-10. The pivot shift test. A palpable or visible pop forward of the tibia during early flexion with a valgus stress implies anterior cruciate ligament laxity.

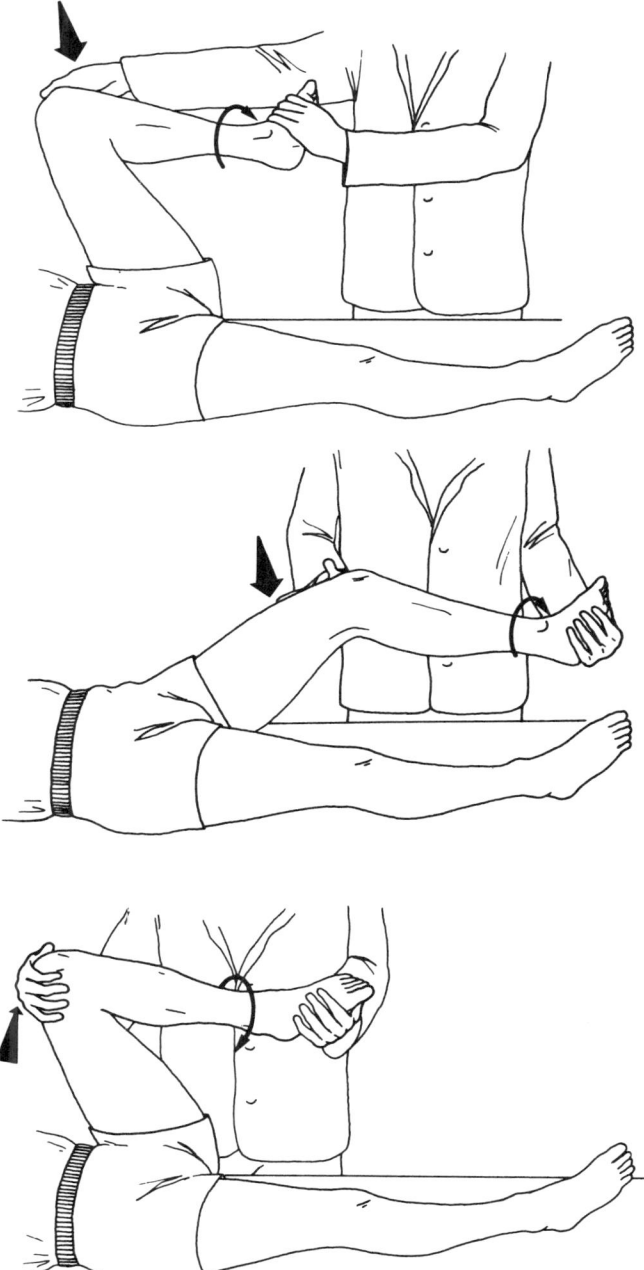

Fig. 15-11. McMurray's test. The top two drawings on this page illustrate medial tears; the bottom one here and the top one on the next page illustrate lateral tears.

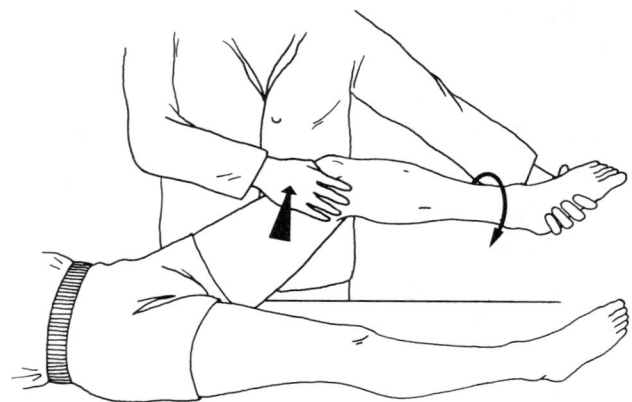

Fig. 15-11. Continued.

in athletes and others with anterior knee pain is the tightness of the hamstring and calf muscles (see Figure 15-13) and the amount of pronation of the foot (see Chapters 18 and 20). Measuring Q-angle (Figure 15-14) is generally only of academic interest, since it is not correctable. Table 15-1 summarizes the steps in examining the knee.

KNEE PAIN

Evaluation

See the discussion of history and examination above.

Vague pain in the anterior knee which is made worse by stairs, squatting or kneeling and by prolonged flexion is suggestive of PATELLOFEMORAL PAIN, which is by far the most common cause of knee pain in teenagers and young adults. OSTEOCHONDRITIS DISSECANS may present with similar symptoms, and RECURRENT PATELLAR SUBLUXATION often appears as an especially severe variant of the patellofemoral pain syndrome.

MENISCUS TEARS can be present without a clear history of specific precipitating trauma, especially in older patients. They are suggested by localized joint line tenderness, a positive McMurray's test or a history of locking or frequent giving way.

The diagnosis of TENDONITIS is usually fairly clear, with localized tenderness at the inflamed tendon.

Most older patients presenting with knee pain have DEGENERATIVE ARTHRITIS as the cause of their discomfort. Rarely an older patient with severe medial knee pain will have OSTEONECROSIS (avascular necrosis) of the medial femoral condyle (Early diagnosis is by bone scan, management by the orthopedist.)

KNEE PAIN

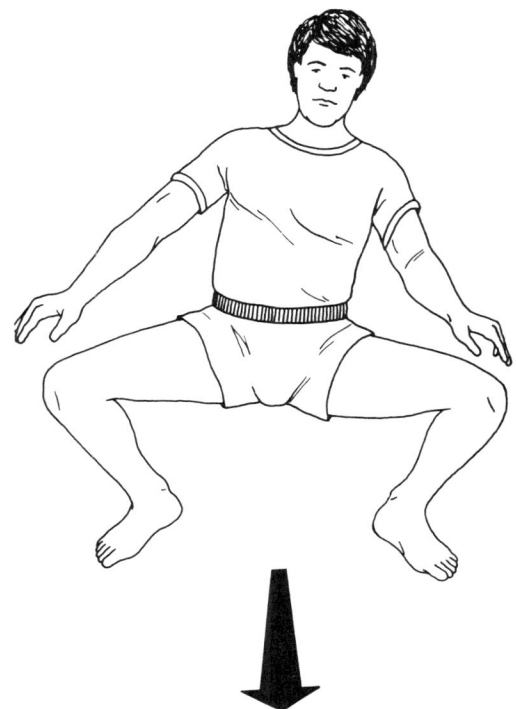

Fig. 15-12. The duck-walk maneuver.

Patellofemoral Pain (patellofemoral dysfunction, extensor mechanism dysfunction, anterior knee pain syndrome, chondromalacia patellae) *(Very common)*

Diagnosis

The patient usually complains of vague aching and often crepitus in the anterior knee. Symptoms are usually worse after squatting or kneeling and with going up or down stairs or hills.

Stiffness after prolonged flexion is often an associated complaint (as in after sitting in a theater: "movie sign").

Vague *posterior* knee pain related to hamstring tightness can be associated as well.

The patient will often complain of episodes of "almost" giving way, but rarely true collapse.

The syndrome is especially common in teenagers and young adults (especially females) and in runners (discussed in Chapter 20).

On examination, there is usually peripatellar tenderness and pain, and crepitus on patellar compression or inhibition testing (Figure 15-6).

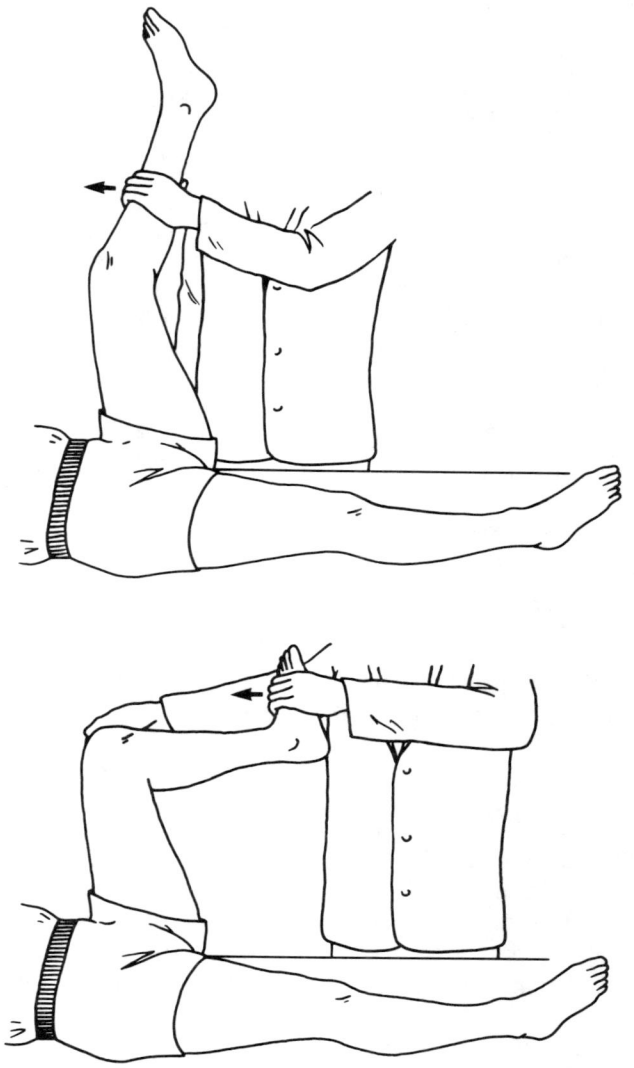

Fig. 15-13. Measuring hamstring (top) and calf tightness (bottom).

KNEE PAIN

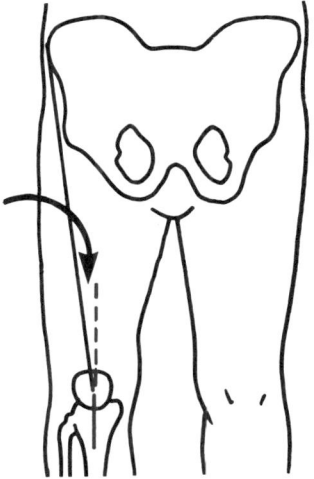

Fig. 15-14. The Q-angle (quadriceps angle) is that formed between a line from the patella to the anterior superior iliac spine, and another from the center of the patella to the tibial tubercle. More than 10° in males or 15° in females is considered excessive. The higher the Q-angle, the more likely is patellofemoral dysfunction.

Unilateral or severe symptoms raise the possibility of OSTEOCHONDRITIS DISSECANS or RECURRENT PATELLAR SUBLUXATION. Finding a tender thickened band just medial to the patella, or a complaint of painful snapping, suggests an INFLAMED SYNOVIAL PLICA as the diagnosis.

Pathophysiology

Previous trauma or degenerative change to the cartilage that lines the joint between the patella and the femur can predispose the area to pain . . . but this is certainly not a universal finding, which is why the name "Chorndro Malacia" ("soft cartilage") has fallen out of favor.

Various abnormalities can affect patellar tracking, i.e. the way in which the kneecap moves during flexion and extension of the knee; many of these (such as an increased Q-angle, high riding patella, etc.) are anatomic factors which are not easily correctable, so not really worth looking for. But relative weakness of the quadriceps, specifically the VMO (which is the main active stabilizer of the patella, and which inserts obliquely into the kneecap and has a different innervation from the rest of the quadriceps) is a critical factor in abnormal patellar tracking, and is correctable. Another potentially correctable factor affecting patellar tracking is excessive pronation of the foot (see Chapter 18).

The compressive force between the patella and the femur is another determinant of stress on this joint. This is increased with increased transmitted shock up the leg (as with hard terrain and inadequately cushioned shoes), and by muscle tightness (especially of the hamstrings and to a lesser extent, the calves).

Whether the patellofemoral pain syndrome is always or sometimes a precursor of degenerative arthritis of the patellofemoral joint is still an open question.

Treatment

Advise the patient to *modify activity* as necessary to reach a tolerable level of symptoms. Kneeling and squatting should be discouraged.

Table 15-1   Examination of the Knee

With the patient lying supine:

1. Measure each thigh circumference 6 cm proximal to the superior pole of the patella. Observe quadriceps and VMO tone.
2. Look for joint effusion or other swelling.
3. Check passive range of motion.
4. Look for sites of tenderness.
5. Check medial and lateral stability with the knee flexed 30° and then with it straight.
6. Do the anterior and posterior drawer tests as well as Lachman's test.
7. Perform the patellar compression and apprehension maneuvers. Observe patellar tracking.
8. Check for hamstring and calf tightness.
9. Do McMurray's test.

Have the patient roll over to the prone position:

10. Look for tenderness or swelling in the popliteal fossa.

Have the patient stand up:

11. Ask the patient to do a deep knee bend on each leg.
12. Have the patient duck-walk.
13. Have the patient hop on each leg.
14. Observe foot pronation.

With some practice the entire examination will take no more than a few minutes. Of course, additional examination should be done if indicated.

Symptomatic treatment such as *aspirin* may be used as needed.

The patient should be placed on a *strengthening* program for the *quadriceps* and VMO. Stationary cycling against high resistance (with the seat raised) and progressive resistance short-arc quadriceps exercises are best.

If there is hamstring or calf tightness, a *stretching* program should be instituted.

Better *cushioned footwear* can be helpful. *Correction* of any excessive foot *pronation* may also be beneficial.

A patellar *brace* (a compressive bandage with a hole cut out for the patella and reinforcement built in to help correct patellar tracking) is sometimes worth a try.

Referral for *surgery* is a last resort. Possible procedures include arthroscopically smoothing out rough spots in the joint, or releasing a tight lateral retinaculum to improve tracking.

Keep in mind that the syndrome very often self-improves in time.

Recurrent Patellar Subluxation *(Uncommon)*

Diagnosis

The symptoms are similar to those of patellofemoral pain (with which it usually coexists), but are often more episodic and severe. There may or may not be a

KNEE PAIN

past history of frank patellar dislocation, and the patient may or may not realize that her patella is subluxing.

On examination, there is usually peripatellar tenderness, marked lateral patellar tracking, and a positive patellar apprehension test.

Pathophysiology

Previous patellar dislocation, tight lateral retinaculum, and weak medial quadriceps are the main causes.

Treatment

Quadriceps strengthening exercises are essential; often the supervision of a physical therapist is best.

A patellar *brace* may be helpful.

Often referral for *surgery* is necessary.

Plica Syndrome *(Uncommon)*

Diagnosis

The symptoms are often similar to those of patellofemoral pain, but often the patient will complain of a painful snapping sensation, and examination may reveal a tender thickened band medial to the patella.

Pathophysiology

A remnant of a developmental "seam" in the synovium becomes inflamed.

Treatment

Surgical release (usually through the arthroscope) is curative.

Breaststrokers Knee *(Common in breaststrokers)*

Diagnosis

A swimmer who does a lot of breaststroke complains of medial knee pain without any history of specific trauma. Examination reveals diffuse medial knee tenderness.

Pathophysiology

Though the specific structure involved has not yet been delineated, the medial knee is placed under stress with repetitive "whip kicking" while doing the breaststroke.

Treatment

Ice and *antiinflammatory medication* may be used. The patient should decrease the amount of breaststroking enough to allow symptoms to resolve. Obtaining the advice of a good swimming coach about *proper mechanics* may help alleviate the problem.

Meniscus Tears (Torn Cartilage) *(Common)*

Diagnosis

The patient often gives a history of twisting the knee with the foot planted, leading to sudden pain in the knee (with or without an associated crunching sensation), followed by swelling several hours later.

Or there may be a history of chronic or recurrent knee pain and recurrent swelling. Note that history of the inciting trauma is often completely lacking in these chronic cases.

The knee may recurrently collapse.

Localized joint line tenderness can be found on examination, and is often the only clue.

Often there is also pain on approach to full flexion.

The patient is usually unable to do a good deep knee bend on the involved knee, and duck walking may lead to pain and/or clicking (Figure 15-10). McMurray's test (Figure 15-11) is diagnostic if positive but are often negative.

Uncommonly a firm lump may be felt over the joint line (the so-called *meniscal cyst*).

It is very common for arthroscopy to be necessary to make or confirm the diagnosis, especially in chronic cases.

A fold of synovial membrane *(plica)* can often mimic a meniscal tear, the correct diagnosis only being made on arthroscopy.

And often, lesions of the articular cartilage of the medial and lateral compartments can be indistinguishable by history and examination from meniscus pathology.

Pathophysiology

The meniscus is caught between the femur and tibia and torn, or torn because of stresses on its various connections. The medial meniscus is involved much more frequently than the lateral.

Older patients in whom the menisci have degenerated to some extent require only very minor trauma to tear them. These "degenerative meniscal tears" can also result from chronic ligamentous instability.

A "discoid" lateral meniscus can mimic a tear, especially in young girls.

KNEE PAIN

Treatment

Acute Tears:

Any case in which an associated ligamentous tear cannot be definitely ruled out should be referred. Otherwise: *Immobilize* the knee for one to several weeks or merely decrease weight-bearing, depending on severity.

The patient should be instructed on *protective* measures (not walking on uneven ground or on high heels, no activities which require sudden starts and stops, no squatting, etc.).

Quadriceps strengthening exercises (Patient Handout 15-1) should be used to prevent atrophy.

Ice and *analgesics* may be prescribed for symptomatic relief.

Meniscal tears sometimes heal over six weeks or so, especially those of the more fibrous outer portion which has a better blood supply, and in many cases no surgery will be necessary.

However, a locked knee (one in which full extension is impossible) must be seen by an orthopedist within 24 hours, and severe tears (associated with much swelling and collapsing) are also best handled by the specialist.

In addition, any case in which an associated ligamentous tear cannot be definitely ruled out should be referred.

Patients Not Presenting With The Acute Episode:

There is increasing evidence that removal of menisci hastens degenerative change in the knee, which is why more and more arthroscopic repairs and partial meniscectomies and less complete meniscectomies are being done. Therefore:

If the pain is severe enough or the disability great enough that the patient would consider surgery, or if the patient's knee is frequently giving way, he should be *referred* to the orthopedist for definitive diagnosis and treatment.

But if those conditions are not present, treat with prn *rest, ice, analgesics/ antiinflammatories,* strength maintenance and avoiding reinjury with option for referral later.

Patellar Tendonitis (Jumper's Knee) *(Common)*

Diagnosis

Tenderness is found over the patellar tendon or its insertion at the tibial tubercle (Figure 15-4). (But if the patient is a teenager, see OSGOOD-SCHLATTER'S DISEASE in Chapter 19.) This is often associated with PATELLOFEMORAL PAIN.

Pathophysiology

Inflammation is due to excessive stress, usually due to sudden contraction of the quadriceps (necessary at the moment of heel strike in jumpers or runners to prevent knee collapse, since the knee is somewhat flexed at that moment). When inflammation is severe or prolonged, sudden rupture is a possibility.

Treatment

    Rest by decreasing or eliminating the responsible activity.
    Use *aspirin* or a course of *NSAIs*.
    Ice may be helpful. So may compression.
    Injection is contraindicated due to the great stresses at this point and risk of subsequent rupture.

Anserine Tendonitis *(Common)*

Diagnosis

    Tenderness is at the anserine tendon insertion (Figure 15-4).

Pathophysiology

    Excessive strain there leads to a tendonitis, or to inflammation of a bursa that develops between tendon and bone proximal to the insertion.

Treatment

    Use *rest, ice,* and *antiinflammatories*.
    Steroid injection has a good response rate. See Figure 15-15.

Degenerative Arthritis of the Knee *(Very Common)*

Diagnosis

    Chronic pain is worse after weight-bearing activity, better with rest.
    Found in older patients and those with previous significant trauma.
    Tenderness may be in the medial joint line (medial compartment), lateral

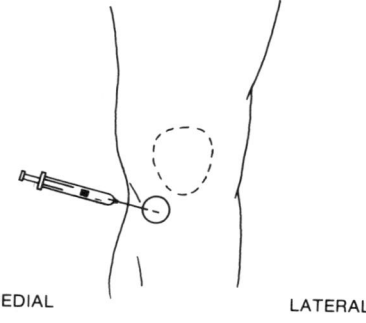

MEDIAL   LATERAL

Fig. 15-15. Injecting the Anserine Tendon Insertion. After prepping the area and spraying with ethyl chloride, use a 23 or 25-gauge needle to infiltrate a cc of steroid and a cc of local anesthetic around the tender area. See pp. 24-26 for further details and precautions.

KNEE GIVING WAY AND CATCHING/LOCKING

joint line (lateral compartment), along the patellar borders with a positive patellar compression test (patellofemoral compartment), or any combination thereof.

Occasionally there is some swelling.

There is a variable amount of pain with range of motion, which can sometimes be restricted.

Loss of muscle tone is often found.

When severe in the medial or lateral compartment, varus or valgus deformity may develop.

The X-ray may be normal early (when changes are limited to cartilage), but some findings are usually present. (But remember that the presence of degenerative changes on X-ray does not prove that any particular symptoms are due to those changes.)

Pathophysiology

Degeneration of articular cartilage.

Treatment

The patient must find a proper *balance* of *rest* and *use*: enough activity to maintain strength and flexibility, but no so much as to increase inflammation.

Aspirin or *NSAI's* are helpful.

Heat to relieve stiffness and *ice* when acutely inflamed should be recommended.

Weight loss is critical if the patient is obese. A *cane* can sometimes be helpful.

When there is quadriceps weakness or loss of range of motion, *strengthening* and *restoring motion* are critical. Often supervision by a physical therapist is best.

Shoes which absorb shock better or inserts for the same purpose often help.

Technique of *injection* is shown in Figure 15-13; it should be reserved for moderately severe exacerbations.

Referral for arthroscopic or open joint debridement, for osteotomy, or for total joint replacement should be considered in severe cases not responding to conservative management, especially those with limb deformity or night pain.

KNEE GIVING WAY AND CATCHING/LOCKING

Evaluation

See the discussion at the beginning of the chapter.

A vague giving way sensation, usually without actual falling, is common in PATELLOFEMORAL PAIN and symptomatic DEGENERATIVE ARTHRITIS.

Giving way suddenly with stepping in certain directions, often followed by a day or two of pain and swelling, is suggestive of RECURRENT PATELLAR SUBLUXATION (especially in teenagers; the displacement of the patella may or may not be noticed) or CHRONIC LIGAMENTOUS INSTABILITY (especially with a history of previous knee injury).

A mild catching sensation is often associated with PATELLOFEMORAL PAIN; but more severe or better defined symptoms, especially if associated with giving way, suggests a MENISCAL TEAR or a loose body ("joint mouse") from OSTEOCHONDRITIS DISSECANS or previous trauma.

A painful snapping sensation is often indicative of an inflamed SYNOVIAL PLICA.

A true locked knee (one which cannot be extended because of a mechanical block) following trauma should be referred immediately to an orthopedist.

Chronic Ligamentous Instability *(Common with a history of prior trauma, rare without)*

Diagnosis

The history is usually of recurrent episodes of giving way when stepping in certain directions, often with a day or two of pain and swelling following each episode.

Symptoms of patellofemoral pain, degenerative arthritis or meniscus tears often coexist.

Careful examination usually reveals instability . . . most commonly anterior or anterolateral (see Figure 15-8). The pivot shift sign (Figure 15-9) is often positive. Other instabilities may be found also, as well as signs of patellofemoral pain syndrome, degenerative arthritis or meniscus tears.

Pathophysiology

The anterior cruciate ligament is usually at the center of the problem. The exact type of instability found is a function of what parts of the anterior cruciate are affected, how badly they are dysfunctioning, and what other structures are also involved.

Long term instability causes synovial irritation, meniscal tears, and premature degenerative arthritis.

Treatment

Intensive *muscular rehabilitation,* preferably under the supervision of a physical therapist, is essential to help provide "dynamic stability" to the knee that lacks ligamentous stability. Note that the hamstrings, not the quadriceps, are the main dynamic stabilizer in the knee that lacks an adequate anterior cruciate ligament. Stationary cycling against resistance is thus an excellent rehabilitative measure for these patients.

Customized *braces,* bulky and expensive, can be helpful; this is best left to the orthopedist, who should also be allowed to evaluate the patient for various *surgical* procedures which may be helpful in alleviating symptoms and retarding degeneration of the knee.

KNEE SWELLING

Osteochondritis Dissecans *(Uncommon)*

Diagnosis

Usually found in teenagers and young adults.
There is sometimes pain or tenderness under the patella.
Often the patient gives a history of knee collapse.
An area of osteochondritis may be visible on a "tunnel view" X-ray; a loose body may be visible if it is calcified.
Tomograms or referral for arthrography or arthroscopy may be necessary to make the diagnosis.

Pathophysiology

A piece of articular cartilage with or without a piece of attached bone becomes necrotic and can break off the femoral surface (most commonly the lateral border of the medial condyle) and becomes a loose body in the joint. This may be due to an anomaly of ossification, localized ischemia, or injury.

Treatment

Refer the patient to an orthopedist for management.

KNEE SWELLING

Evaluation

Swelling of the prepatellar bursa or the infrapatellar bursa must be differentiated from swelling of the knee joint itself (Figure 15-2). The latter implies an intraarticular process; if acute and atraumatic, leading possibilities are infection, gout, and pseudogout. (The last-mentioned is especially common in the knee joint; a clue to its presence is an X-ray showing calcium deposits in the menisci or articular cartilage. Pseudogout is discussed in Chapter 24.) Swelling within the joint that occurs within minutes of injury implies blood in the joint, and indicates a fracture, patellar dislocation or cruciate ligament tear until proven otherwise. Swelling occurring hours after injury or recurrent swelling may be due to any irritated mechanical lesion (DA, torn meniscus, etc.) or of course to inflammatory arthritis. Swelling in the posterior aspect of the joint is the so-called BAKER'S CYST.

Prepatellar Bursitis (Housemaid's Knee) *(Common)*

Diagnosis

A painless or slightly painful swelling is localized over the inferior patella and the patellar tendon (Figure 15-2).

If there is redness, warmth, significant tenderness, or a nearby break in the skin, it *must be tapped* to rule out infection.

Pathophysiology

Fluid accumulates in the inflamed bursa usually from prolonged kneeling.

Treatment

The patient should *avoid kneeling*; a *pad* may be used to prevent recurrence if kneeling is unavoidable.

Heat may help.

Drainage may be offered; the fluid usually reaccumulates quickly unless some steroid is instilled after drainage, but even this often does not work. The method is simply to enter the center of the swelling with a .#18 needle (with a large syringe) after sterile preparation and ethyl chloride spray; never inject steroid if the fluid drained appears cloudy or purulent. Remember, the bursa must be aspirated if the signs of infection described above are present. A compressive bandage should be worn afterwards.

Surgery may be done to remove a chronically or recurrently inflamed bursa.

Infrapatellar Bursitis *(Uncommon)*

Diagnosis

A painless or slightly painful swelling is found as shown in Figure 15-2.

Pathophysiology

The bursa becomes inflamed from overuse (flexion/extension) of the knee.

Treatment

Rest, heat, and *antiinflammatories* may be helpful.

Baker's Cyst *(Uncommon)*

Diagnosis

A painless or slightly painful swelling occurs in the popliteal space (Figure 15-2).

Be sure it is not pulsatile or associated with a bruit, since a popliteal artery aneurysm can be misdiagnosed; and be sure it is cystic, because soft tissue tumors can present in this location.

Pathophysiology

This may actually be a bursa associated with the semimembranosus or gastrocnemius, which may or may not be in communication with the joint space; or it can be a posterior extension of the synovial space of the knee.

Swelling of the latter type is associated with intraarticular knee problems.

The cyst can occasionally rupture, leading to a swollen inflamed calf which may mimic deep vein thrombosis.

Treatment

Treatment of the *underlying knee problem* is the best treatment of the cyst.

Surgical excision should be resorted to only if the above is not successful and the cyst is sufficiently large and symptomatic.

KNEE TRAUMA

Evaluation:

Always examine distal neurovascular status, and always X-ray.

Be aware tht a dislocated knee (not patella), though rare, is an extreme emergency because of the potential for arterial injury.

Otherwise the prime function of evaluation is to rule out fracture or significant ligamentous injury. If you can't be sure about instability, because of guarding, spasm, swelling or whatever, orthopedic evaluation within a couple of days is necessary.

See the discussion of history and examination in the beginning of this chapter.

Some specific points with regard to history: a "pop" heard or felt at the time of injury is, even without other findings, highly suggestive of an ANTERIOR CRUCIATE LIGAMENT INJURY. (A "tearing" sensation is more consistent with COLLATERAL LIGAMENT INJURY, and a "crunch" with MENISCUS TEARS or ACUTE PATELLAR DISLOCATION.)

Immediate swelling following the injury implies blood in the joint, indicative of a FRACTURE or ACL INJURY (or occasionally an ACUTE PATELLAR DISLOCATION). This history alone is enough to mandate immediate orthopedic referral. (Delayed swelling is usually due to synovial irritation and is not as critical.)

Collateral Ligament Sprain (knee sprain, twisted knee)
(Very common)

Diagnosis

The patient usually reports a twisting type injury to the knee, or externally applied trauma.

For examination see the beginning of the chapter.

Tenderness will usually be found somewhere along the course of the involved collateral ligament. (The medial is more commonly injured than the lateral).

Soft tissue swelling over the injured area is common, but a joint effusion indicates associated injury to other structures or very severe injury. Some ecchymosis may be present.

Tests for stability of the injured and other ligaments are mandatory. (Figures 15-7, 15-8, and 15-9).

Treatment

Mild sprains, in which instability is absent or definitely mild, are treated by: *immobilizing* the knee in a splint and using *crutches* if pain and limp are significant; progessive increase in weight bearing as symptoms improve; frequent *ice* application for the first forty-eight hours; a *compressive bandage*; pain medication if needed; *isometric quadriceps exercises*; *precautions* to avoid reinjury; and slow, gradual resumption of activity and muscle strengthening as improvement occurs.

Cases in which there is significant instability of the ligament, or in which it can't be ruled out because of inadequate examination, or in which associated cruciate ligament injury is suggested or can't be ruled out, need to be protected with a splint and crutches and seen by an orthopedist within a few days.

Occasionally a meniscal tear will coexist; this is often difficult to diagnose acutely, and the patient should be advised of its possibility and that persistence of symptoms may require reevaluation later.

Anterior Cruciate Ligament Injury *(Uncommon, but also underdiagnosed)*

Diagnosis

The injury may occur when coming down hard on the slightly flexed knee, or with a twist, or with externally applied force.

The patient will often report a "pop" at the time of injury and/or immediate swelling, but these are not necessary for the diagnosis.

If Lachman's test (Figure 15-10) or the anterior drawer test (Figure 15-8) reveal instability, or if you can't be sure for any reason, evaluation by an orthopedist is necessary. (He or she can examine the knee under anesthesia or use a device called an arthrometer to determine stability, or sometimes will do diagnostic arthroscopy solely on the basis of history.)

Pathophysiology

Be aware that there is often associated collateral ligament, capsular or cartilage injury.

This injury may lead to chronic ligamentous instability, meniscus lesions and degenerative arthritis.

It is hard to correlate the degree of injury to the ligament to the subsequent type and amount of instability; resulting dysfunction depends on muscle tone, other injured structures, and underlying joint laxity.

KNEE TRAUMA

Some authorities go so far as to describe a knee which is "ACL deficient" as facing the "beginning of the end," but the entire topic is still quite controversial.

Treatment

The decision as to whether to acutely repair the injured ligament or treat conservatively depends on many factors and is left to the orthopedist. The success of late reconstructions has not been good.

Posterior Cruciate Ligament Injury *(Rare)*

Diagnosis

This is usually associated with a severe injury, and the need for orthopedic consultation will not be in doubt.

The posterior drawer test may be positive, and varus/valgus instability in full extension may be noted.

Treatment

By the specialist.

Acute Patellar Dislocation *(Uncommon)*

Diagnosis

There is usually a history of a twist with a grinding/tearing sensation and much swelling. The patella may relocate before it is noted to be out of place.

Pathophysiology

The patella dislocates laterally.

Treatment

Immobilization for several weeks for the first episode, and *muscle rehabilitation*. *Referral* to an orthopedist is recommended.

For the management of recurrent episodes see RECURRENT PATELLAR SUBLUXATION above.

Patellar Contusion *(Common)*

Diagnosis

The patient reports striking the anterior knee, as in a fall forward, without a twist. Effusion may develop, but is not immediate.

Examination reveals patellar tenderness and no suggestion of ligamentous instability. X-rays should be done to rule out fracture.

Pathophysiology

Injury to the patella or the cartilage forming the patellofemoral joint.

Treatment

Symptomatic. Future patellofemoral pain is common.

PATIENT HANDOUT 15-1

Patellofemoral Pain

If your doctor has told you that your aching knee is because of "patellofemoral pain," you've got lots of company. This is the most common cause of knee pain, especially in young people. And though uncomfortable, rest assured that it is usually not serious.

Why does it hurt?

The pain comes from the joint under the kneecap ("patella") where it slides up and down over the thigh bone ("femur"). But *why* it happens, no one knows for sure. We do know, however, some factors that contribute to it. Some you can't do anything about . . . the anatomy of your legs and kneecaps, and whether you have fallen on or banged up your knees in the past, for instance. But some you can affect: mainly the strength and flexibility of your muscles.

What should I do about it?

First of all, understand that if that's what you've got, its *not* serious.

Second, try to *avoid kneeling and squatting* as much as you can. Don't wear a compressive bandage unless your doctor recommends one.

Third, use *aspirin* or put *ice* on it (for twenty minutes at a time) if and when you need to.

Fourth, *strengthen your quadricep muscles.* These are the large muscles in the front of your thigh. They attach to the top of your kneecap and help it move properly. You can do this by pedalling a stationary cycle each day, with the resistance turned up (and the seat raised to your knee doesn't bend too much.) Or you can do the exercise shown.

If the muscles in the back of your leg are tight, stretching them may help too. Do it after your muscles are warmed up, and hold each stretch for a minute almost tight enough to cause pain but not quite. *Don't bounce.*

It may also help to wear shoes that are more cushioned, or to correct hyperpronating feet if your doctor thinks that's part of the problem. (You will be told how.)

Good luck!

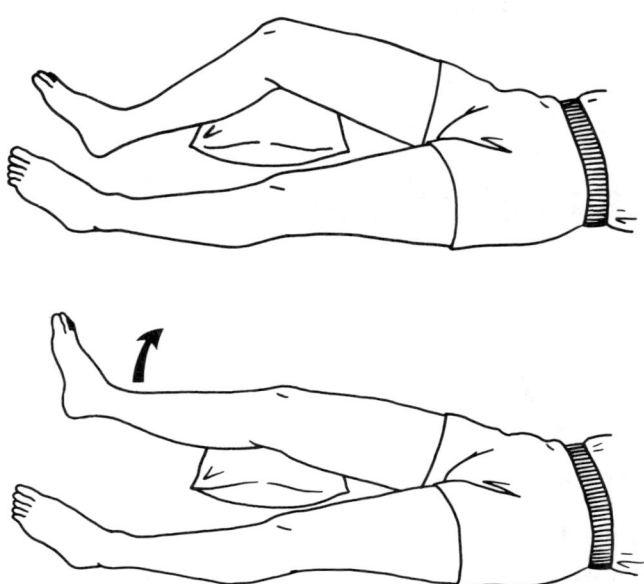

Have your knee bent about 30° over a pillow. *Slowly* straighten your leg, then slowly lift the whole leg. Allow it to go back down slowly. Repeat ten times, rest, ten times more, rest, and try for ten times more. If you can finish all thirty, add weights to your ankle the next day (a handbag with something in it works well). Keep gradually increasing the weight as you get stronger. Do both legs. It's also a good idea to turn over on your stomach when you are done and strengthen your hamstrings too, by bending your knee and letting it drop repetitively with the weight on your ankle.

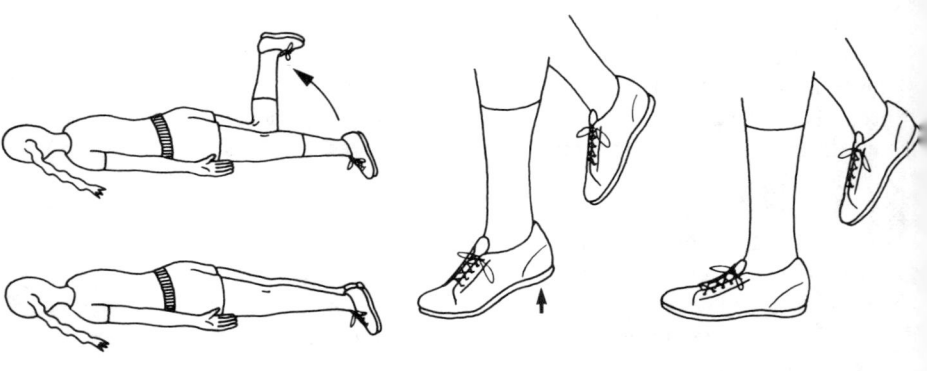

HAMSTRING STRETCH   CALF STRETCH

PATIENT HANDOUT 15-2

Knee Exercises

Do the exercises your doctor recommends.

QUADRICEPS EXERCISES

Quad Sets

While standing, slowly tighten your thigh muscles (pulling your kneecap upward) and keep them tight for a count of ten; then slowly relax. Do this at least ten times an hour whenever you are standing. After a while you will be doing it automatically, without even having to think about it.

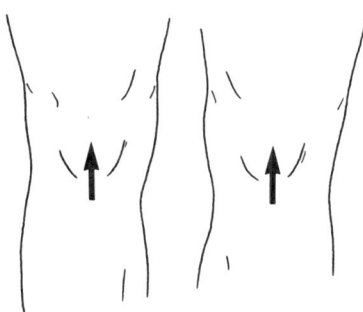

Crossed Leg Isometrics

Either while sitting or lying down, place the heel of one foot in front of the ankle of the other, and push against each other, hard, for a count of five. Repeat ten times, then switch legs.

Straight Leg Raises

While lying on your back on the bed or floor, slowly raise your leg about a foot off the surface, keeping your knee straight. Hold it for a count of five and then slowly put it down. Do 10 to 15 repetitions twice a day. As you get stronger, you may even want to put weights around your ankle (books in a handbag do quite nicely). Athletes can lift about a quarter of their body weight during these exercises.

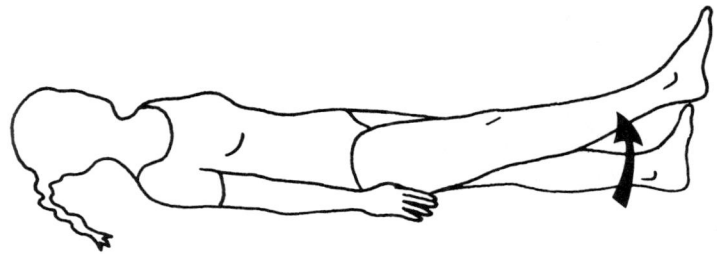

Short Arc Quadriceps Progressive Resistance Exercises

Do this like the straight leg raises above, but instead of having your legs straight, start with it bent over a pillow, about 30°. Slowly straighten your leg, then slowly lift it, slowly let it down and then let it bend. Two or three sets of ten in a row, with weight on your ankle increasing as you get stronger.

HAMSTRING EXERCISES

While lying on your stomach, raise your foot (by bending your knee) slowly into the air and hold for a slow count of five. Then slowly put it back down. Do ten to fifteen repetitions twice a day. As you get stronger, add weights around your ankle; aim for fifteen percent of your body weight.

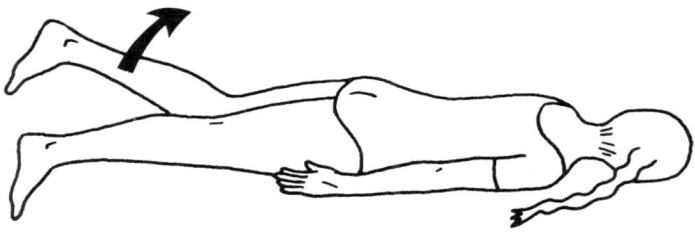

Working on a stationary cycle against turned-up resistance is a good way to strengthen both your quadriceps and your hamstrings.

CALF EXERCISES

ABDUCTOR EXERCISES

While lying on your side (with the affected leg up) raise your leg (at the hip) about a foot into the air and hold for a slow count of five. Then slowly put it back down. Do ten to fifteen repetitions twice a day.

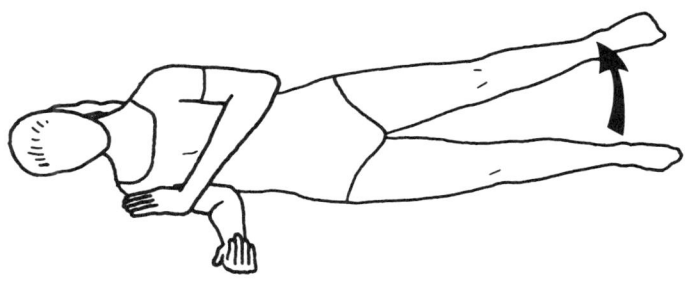

CALF EXERCISES

Hold on for balance, stand on one foot and go up on your toes as many times as you can. If you get past thirty, do it with weights in your hands the next day, and gradually increase the weight as you get stronger.

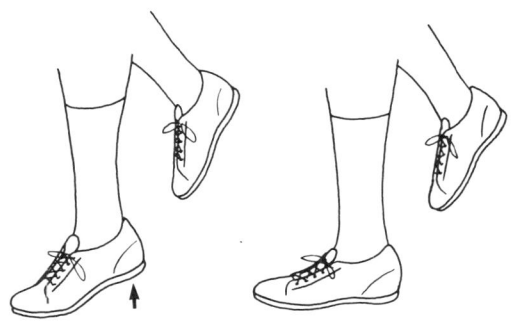

PATIENT HANDOUT 15-3

Ski Safety

Downhill skiing can be fun and exhilarating . . . and dangerous. Out of every one hundred and fifty skiers on the slopes on any day, one will end his or her day in the hospital or clinic, sometimes with a serious injury. But it doesn't have to happen to you. With proper attention to your equipment and a little common sense, you can greatly decrease the chance of a serious injury.

Probably the single most important factor is the *binding*. That's what goes between the boot and the ski, and often it's what is between you and a trip to the emergency room. Simply put, a binding is supposed to hold your boot to the ski but also to release before the ski twists or bends the leg enough to cause damage. What kind of binding should you choose? This isn't the place to skimp on quality, that's for sure. You should pick a binding that releases in as many directions as possible, not just up at the heel and twisting at the toe . . . because while most skiing falls happen that way, not all of them do. But just as important is setting and maintaining them properly. A ski shop can advise on the proper setting, depending on weight, bone size and ability level. A beginner should set it even lighter than recommended, because he or she will be falling a lot. A binding that comes off when it shouldn't is usually just an inconvenience; a binding that doesn't come off when it should can be a disaster. Bindings must not *ever* be set too high.

Maintaining them properly is very important. Bindings need to be kept clear of accumulated snow and ice, especially over the "anti-friction device" toward the front. Before every single day of skiing, they should be tested by purposely releasing them in all directions. And lubricating them once a year is a good idea too.

Other equipment affects safety too. Ski brakes are much safer than retention straps. If you fall with your skis flailing around by the straps, you can be badly cut by them, often around the head or face. Also, the poles should be held without straps, because a fall forward with the pole held there commonly results in a very serious sprain to the inside of the thumb.

Choose skis and boots properly. Skis that are too long for your ability increase the chance of trouble. And boots that are too loose or left partly unbuckled are an invitation to ankle injuries.

There are nonequipment factors, too. Conditioning and strengthening your muscles before ski season begins helps prevent trouble. Your doctor can give you some exercises to do.

Two other very important factors are: not skiing beyond your ability, and most critical of all, not skiing when fatigued. That's when the great majority of injuries occur. That can't be emphasized enough.

A few other tips: don't ignore knee or thumb sprains (they can be more serious than you think at the time); take just as much care with your kids' equipment as with your own; and, as a final thought, drive carefully and wear your safety belts on the way to the slopes.

Injuries can still occur, but if you follow this advice you will be much more likely to stay out of trouble.

16

The Lower Leg

Calf Pain 206
    Evaluation 206
    Calf Pull 207

Neurologic symptoms and *diffuse pain of the lower leg* are discussed in Chapter 13. Because *shin pain* is almost invariably a complaint of athletes, discussion is deferred to Chapter 20. *Cramps* are covered in Chapter 21.

CALF PAIN

Evaluation

Calf pain can of course be due to *Deep Vein Thrombosis,* a potentially life threatening condition. The diagnosis of this condition is beyond the sope of this book (and in fact is quite difficult to make without performing invasive tests). Its possibility must never be forgotten, and history, examination and tests appropriately performed to rule it out before one of the conditions discussed in this section is diagnosed.

A ruptured BAKER'S CYST is another cause of acute calf pain and swelling that must be considered.

Calf pain that comes on with activity and is relieved with rest in a fairly reproducible pattern is suggestive of *Claudication,* and appropriate examination and tests must be performed to investigate this possibility.

Ask about any specific *trauma, strain* or *overuse.*

CALF PAIN

Calf Pull *(Common)*

Diagnosis

    The patient gives a history of acute or repetitive strain.
    Pain and tenderness can be located anywhere in the calf.
    Deep vein thrombosis must be ruled out.
    Check vascular status.

Treatment

    Use *ice* in the acute situation.
    Apply *compression*.
    Advise *rest* (crutches if severe) with resumption of activity as tolerated.
    Antiinflammatory medications may hasten resolution if the pull is severe.
    A *heel lift* may be of temporary symptomatic benefit.
    To prevent recurrence, *strengthening* and *stretching exercises* should be prescribed. See Patient Handouts 15-3 and 20-2.

17

The Ankle

Ankle Pain 210
    History 210
    Examination 211
    Posterior Tibialis Tendonitis 212
    Peroneal Tendonitis 213
    Anterior Tibialis and Extensor Tendonitis 214
    Degenerative Arthritis of the Subtalar Joint 214
    Degenerative Arthritis of the Ankle 215
Ankle Sprain 216
    Examination 216
    X-Ray 217
    Ankle Sprain 217
Ankle Instability and Locking 219
    Evaluation 219
    Chronic Ankle Instability 220
Ankle Lumps 220
Achilles Tendon Area Pain and Trauma 220
    Evaluation 220
    Posterior Calcaneal Bursitis 221
    Retrocalcaneal Bursitis 221
    Achilles Tendonitis 221
    Achilles Strain 223

Problems in the inferior *heel* are covered in Chapter 18. Ankle and heel cord pain in *children* are covered in Chapter 19. Ankle pain in *runners* is discussed in Chapter 20. For further details on therapeutic suggestions made in this chapter, see Chapter 3.

    For the anatomy of the ankle see Figure 17-1.

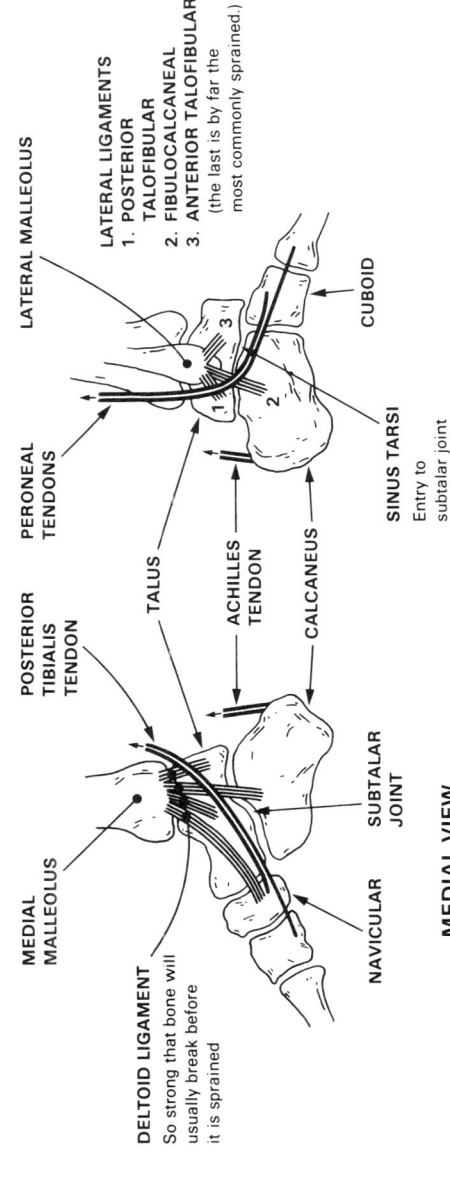

Fig. 17-1. Anatomy of the ankle.

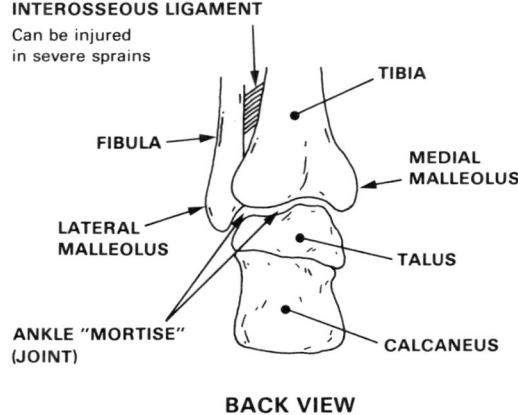

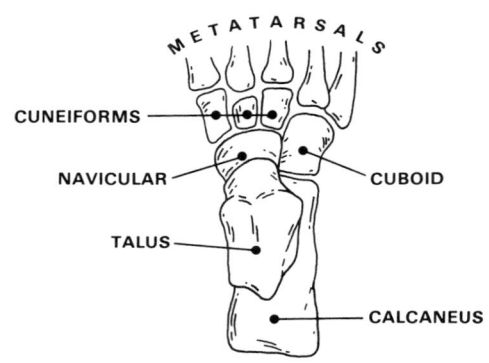

Fig. 17-1. Continued.

ANKLE PAIN

History

    Has there been any specific *trauma, strain* or *overuse*?
    What is the *duration* and the *location* of pain?
    Does the ankle *twist recurrently*?
    What are *precipitating* and *relieving* factors?
    Ask about any *previous* ankle problems, diagnoses or treatments.
    Are any *other joints* involved? Is there any history of arthritis or *gout*?

ANKLE PAIN

Examination

Localize *tenderness* (Figure 17-2).

Any generalized *joint swelling* in the absence of trauma implies an intra-articular process (which if acute must be tapped to rule out infection: see Figure 17-3). This must not be confused with the much more common finding of edema in the soft tissues secondary to venous insufficiency, congestive heart failure and other conditions. See Figure 17-4.

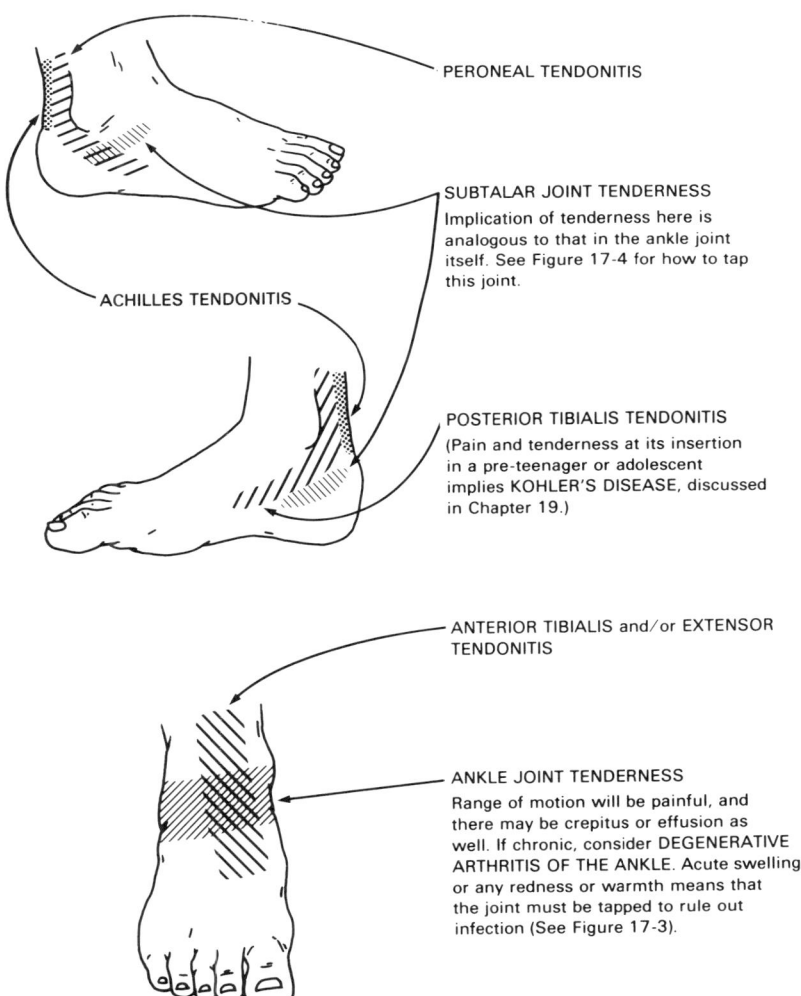

Fig. 17-2. Sites of tenderness: the ankle.

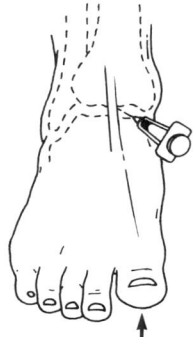

Fig. 17-3. How to aspirate or inject an ankle joint. Find the large extensor hallucis longus tendon in the front of the ankle (have the patient dorsiflex the big toe to make it stand out), and then find the joint line (between the tibia and the talus) just medial to it. Prepare the site and use sterile technique throughout. Inject some anesthetic in the direction shown, then withdraw and insert a 20-gauge or 18-gauge needle on a 20-cc syringe in the same direction, and aspirate. If steroid is to be injected, use a hemostat to hold the needle bevel and change syringes to one containing 3 to 6 cc of steroid. (Never inject steroid if the synovial fluid is cloudy or if there is any other reason to suspect infection.) Steroid can be injected in the same way, without aspiration, using a 22-gauge needle. Be sure to see pp. 13 and 24 for discussion of synovial fluid analysis and of precautions and risks of steroid instillation.

Any *redness* or *warmth* of course indicates an acute inflammatory process.

Test *range of motion*. Normal is 20° dorsiflexion, 45° plantar flexion, 30° inversion, and 20° eversion. Note that the last two numbers include some motion in the subtalar joint.

Test sensation, strength, pulses and distal capillary filling.

Posterior Tibialis Tendonitis *(Common)*

Diagnosis

Tenderness is found at the tendon's insertion on the tarsal navicular or first cuneiform bone and/or along its course behind the medial malleolus (Figure 17-2).

Pathophysiology

Excessive stretch is due to excessive foot pronation.

Sometimes the tendon inserts on an *accessory navicular bone,* and the pseudoarthrosis between the accessory and the true navicular becomes painful.

The problem may rarely be a residual inflammation after trauma, or be due to pressure from ill-fitting shoes.

The adjacent flexor tendons may also be inflamed.

ANKLE PAIN

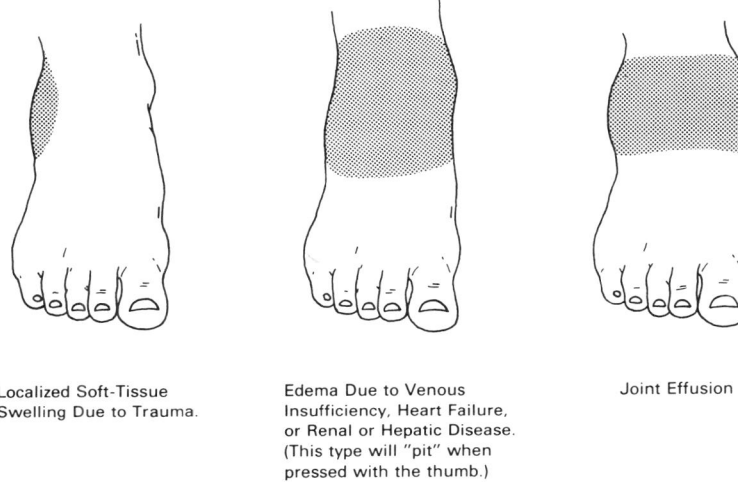

Localized Soft-Tissue Swelling Due to Trauma.

Edema Due to Venous Insufficiency, Heart Failure, or Renal or Hepatic Disease. (This type will "pit" when pressed with the thumb.)

Joint Effusion

Fig. 17-4. Swelling around the ankle.

Treatment

Correct excessive pronation (see Chapter 18).

The sore area should be *padded* to reduce pressure.

In addition, a course of oral *antiinflammatory medication* may be helpful.

In prolonged or severe cases, *steroid injection* may be offered. Success rate is good if abnormal mechanics are corrected as discussed above. Technique is simply to inject (after sterile prep and ethyl chloride spray) a cc or so of steroid and a cc or so of local anesthetic around the painful area (*never* into the tendon) using a 25-gauge needle. See Chapter 3 for further details and precautions before injecting.

Surgery may be resorted to in the presence of an accessory navicular.

Peroneal Tendonitis *(Uncommon)*

Diagnosis

Tenderness is at the peroneus brevis tendon's insertion at the base of the fifth metatarsal and/or along the course of the peroneal tendons behind the lateral malleolus (Figure 17-2).

Pathophysiology

This often follows a typical inversion sprain of the ankle.

Or it can be a result of excessive stretch in patients with a high arched foot and too much supination, or due to local pressure.

Rarely the tendon may be subluxing out of its groove.

Treatment

A *lateral heel wedge* can help.

Oral *antiinflammatories* may be used.

Steroid injection can be offered; success rate is good. The procedure is analogous to that for posterior tibialis tendonitis, discussed above.

Surgery may be considered in recurrent subluxation with severe symptoms.

Anterior Tibialis and Extensor Tendonitis *(Uncommon)*

Diagnosis

Tenderness occurs along these tendons in the anterior ankle and dorsum of the foot (Figure 17-2).

Be sure you are not dealing with a rare infectious tenosynovitis (distinguished by redness, warmth and acute onset).

Pathophysiology

Most commonly this is a pressure phenomenon, resulting from wearing shoes that are too small or tied too tightly.

Sometimes strain is excessive because too-high heels are worn, leading to too much contraction of these muscles (to prevent a slapping gait) and to stretch of the tendons.

Or it can sometimes be posttraumatic (following foot or ankle sprain).

The extensor hallucis longus tendon popping over the anterior talus is the most common cause of the complaint of "snapping ankle." The diagnosis can be confirmed by palpating there as the ankle is actively circumducted.

Treatment

Local *pressure* must be *alleviated* or the heel *lowered*, depending on the suspected mechanism.

Oral *antiinflammatory medications* may be used.

Steroid injection may be offered; success rate is good. The procedure is analogous to that for posterior tibialis tendonitis, discussed above.

Degenerative Arthritis (Osteoarthritis) of the Subtalar Joint *(Uncommon)*

Diagnosis

An older patient or one with significant foot pronation or previous trauma complains of chronic pain which is worse after much weight-bearing.

ANKLE PAIN

Tenderness is found over the sinus tarsi and the subtalar joint line (Figure 17-2).

Often passive range of motion of the joint is painful, and there sometimes is crepitus.

Changes are often seen on X-ray, though it may appear normal early in the course.

Pathophysiology

Excessive pronation or ground reaction forces traumatize this joint between the talus and calcaneus and lead to accelerated degeneration.

See also Chapter 22.

Treatment

The measures recommended in Chapter 18 for patients with chronic foot pain must be employed.

Oral *antiinflammatory* medication may be of benefit.

Injection of *steroid* into the joint may be offered. See Figure 17-5.

Degenerative Arthritis (Osteoarthritis) of the Ankle
(Common)

Diagnosis

An older patient or one with previous trauma will complain of chronic pain which is worse after weight-bearing.

There may be variable diffuse tenderness, and variable pain with range of motion.

Slight joint swelling can occur in an acute exacerbation, but if the joint is red or warm it must be tapped to rule out infection.

The X-ray may be normal early but is characteristic later.

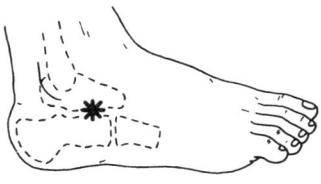

LATERAL

Fig. 17-5. How to aspirate or inject a subtalar joint. Passively invert the ankle, and palpate about a centimeter in front of and just inferior to the tip of the lateral malleolus. The depression felt there is the *sinus tarsi*, through which the subtalar joint can most easily be entered. The rest of the technique is exactly as described in Figure 17-3 (including the cautions *re* infection and steroid instillation).

Pathophysiology

See Chapter 22.

Treatment

Balanced rest and use, heat or *ice,* and *antiinflammatory medications* should be offered.

If the patient is obese, *weight loss* is helpful. Also, adequate shoe cushioning is essential.

Technique of *injection* is shown in Figure 17-3.

Recurrent sprains should be managed as under CHRONIC ANKLE INSTABILITY below.

Consider referral for surgery in the presence of locking or severe symptoms.

ANKLE SPRAIN

Examination

Look for *tenderness*. It is usually limited to the anterior talofibular ligament in first-degree sprains, but often extends to the fibulocalcaneal ligamnet and even the posterior talofibular ligament in more severe sprains (Figure 17-6).

Isolated tenderness over *only* the posterior aspect of the lateral ankle may be indicative of a traumatic subluxation of the peroneal tendons rather than a ligamentous sprain. The tendons have usually spontaneously relocated by the time the patient is seen. The patient is best referred to an orthopedist for management, which may consist of padding, casting or surgery.

If there is significant tenderness and/or swelling over the *medial aspect* of the ankle (deltoid ligament) or the anterior aspect of the ankle, *severe injury* can be assumed and the patient is best referred to an orthopedist.

Soft tissue swelling is localized in first-degree sprains and is more generalized in higher-degree injuries.

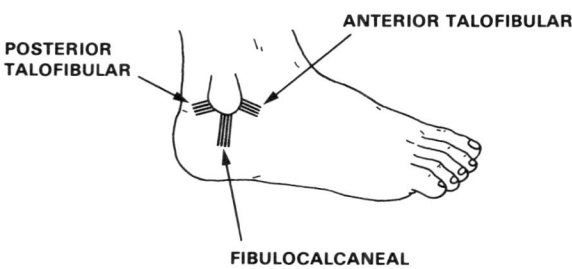

Fig. 17-6. The lateral ligaments of the ankle.

ANKLE SPRAIN

The presence of *ecchymosis* signifies at least some ligamentous tearing, and thus a second-degree injury at the minimum.

Test for *stability* by performing the anterior drawer maneuver (Figure 17-7); good muscular relaxation is necessary for accuracy, and results must always be compared with the uninjured limb because there is great variation from individual to individual. Significant asymmetric laxity implies a third-degree injury, new or old.

Check *range of motion*: dorsiflexion, plantar flexion, eversion and inversion, and compare to the other leg.

Examine the achilles tendon. Significant tenderness, or swelling, or a gap, or lack of ankle plantar flexion when the relaxed calf is squeezed, are grounds for immediate orthopedic referral for management.

See if lateral *compression* of the *distal tibia* and *fibula* is painful. If so, the patient has an injury to the distal interosseous ligament and should be referred to an orthopedist for treatment. Test distal *strength, sensation,* and *capillary filling*.

X-Ray

One should be obtained in all but the most minor injuries.
It will of course be normal in a sprain of the lateral ligaments. (A small

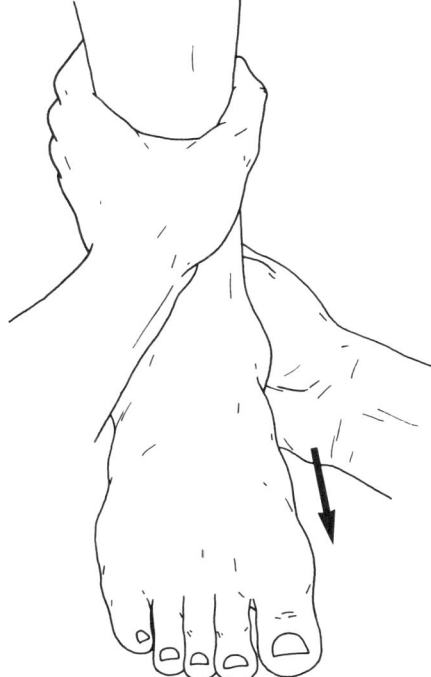

Fig. 17-7. The anterior drawer test (in the ankle). Good muscle relaxation is essential. Always compare to the uninjured side.

avulsion fracture implies total ligamentous disruption and therefore should be treated as a third-degree sprain. Any other fracture is considered beyond the scope of this text.)

If on the AP film a clear space is visible at all points between the tibia and fibula, the patient may well have a disruption of the distal interosseous ligament. If a view of the uninjured leg confirms asymmetry, the patient should be seen by an orthopedist (Figure 17-8).

Since even third-degree sprains are now generally treated conservatively (except in athletes) and since asymmetry in anterior distraction of the ankle is a more reliable sign of ligamentous disruption than excessive inversion, the taking of stress inversion X-rays is usually not necessary. With severe sprains or those in which a joint effusion is found be careful to exclude a fracture of the talar dome.

Ankle Sprain *(Extremely common)*

Diagnosis

The patient usually reports an inversion-type injury. Sometimes the mechanism of injury is not recalled.

On examination, tenderness and soft-tissue swelling are found over the lateral ligaments. Range of motion may be restricted depending on the amount of swelling and guarding.

Ecchymosis or avulsion fracture on X-ray implies at least a second degree injury (partial tear). New instability suggests a third degree injury.

Pathophysiology

Ligamentous injury. A high-arched rigid foot predisposes.

Treatment

Use *ice* for 20 minutes at a time frequently for the first 48 hours, and as needed thereafter. *Never apply heat.*

Apply a *compressive bandage* for as long as there is swelling. An elastic anklet with a horseshoe-shaped piece of foam below the malleolus works well. With severe swelling a boot or Jones dressing may be considered for the early days after injury.

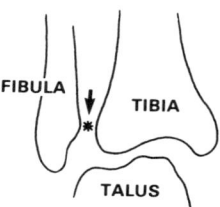

Fig. 17-8. Interosseous ligament disruption. Clear space at all points between the tibia and fibula on the A-P X-ray indicates a possible disruption of the interosseous ligament.

ANKLE INSTABILITY AND LOCKING

Advise strict *elevation* initially, later as needed to control swelling.

If the patient cannot walk with a normal heel-toe gait, *crutches* should be issued and the patient made non-weight bearing until a normal gait (even though still sore) is possible. The amount of walking should be slowly advanced by the patient depending on progress.

Begin dorsiflexion-plantar flexion *exercises* on the second day and circumduction or alphabet exercises soon thereafter. This helps pump out edema and maintain range of motion.

Contrast baths or whirlpool (keep the ankle moving!) can be helpful after the acute phase.

For athletic patients, when walking is essentially pain free the patient may try slow jogging and advance through fast straight running, turns etc. as symptomatic progress allows. *Taping* prior to athletic or dance activity is strongly recommended.

All patients should be advised to exercise *caution* with regards to uneven surfaces, high heeled shoes, etc.

Prolonged cases should be referred to a physical therapist. Also reconsider (and re-X-ray if necessary) the possibility of a talar dome fracture, especially with joint line tenderness or effusion; of a posterior talar process fracture if tenderness is between the talus and achilles*; or a post-traumatic tendonitis, especially if tenderness is found over a tendon (treat appropriately). Prolonged symptoms can also be due to anatomic instability, or inadequate rehabilitation due to lack of patient effort or unrecognized nerve trauma (leading to functional instability).

Note: some patients, especially the elderly or those who cannot or will not use crutches and who have significant injury, will be better off in a short-leg walking *cast* for two to four weeks. Many authorities recommend a cast for all unstable injuries. Be aware that immobilization means a trade-off between ligamentous healing and muscle atrophy, stiffness, and loss of joint proprioception. If cast treatment is chosen, it should not be applied until the acute swelling has subsided, and rehabilitation after the cast is removed is mandatory.

Also note that some authorities feel that *surgical repair* of unstable injuries is indicated in the athlete.

See Patient Handout 17-1.

ANKLE INSTABILITY AND LOCKING

Evaluation

Locking or catching suggests a loose body within the joint from previous trauma or osteochrondritis dissecans of the talar dome, and suggests the need for orthopedic referral.

---

*Hard to differentiate from a common accessory ossicle, the *os trigonum*, on x-ray; bone scan may be necessary.

Instability, or frequent recurrent sprains, is much more common.

Chronic Ankle Instability *(Common)*

Diagnosis

The patient complains of frequent ankle sprains.

A positive anterior drawer sign (Figure 17-7) or inversion instability may be found on examination.

Pathophysiology

This is usually the result of previous trauma, which may have left the ankle with inadequate ligamentous stability, poor joint proprioception, or muscular weakness.

Treatment

The patient should as much as possible *avoid uneven and tilted surfaces*.

Shoes with *high tops* can be helpful. High heels are out of the question. If the patient wears cleats at play, they should be as widely spaced and numerous as possible.

Wearing an *elastic bandage* on the ankle may help to improve joint proprioception and remind the patient of his limitations; it of course does not provide significant mechanical stability.

Using small *lateral heel wedges* is often helpful (Figure 18-3) but can lead to hyperpronation or eversion instability if too large.

A *rehabilitation* program should include exercises to: improve strength, especially of the peroneals; increase flexibility, especially of the calf and achilles; and improve joint proprioception. Supervision by a physical therapist is recommended.

Taping the ankle prior to athletic activity can be quite helpful. Various commercial straps and splints may be more convenient but not as effective.

In recalcitrant cases consider referral to an orthopedist for possible reconstructive *surgery*.

See Patient Handout 17-2.

ANKLE LUMPS

GANGLIA are common about the ankle; see the discussion in Chapter 8, which is equally applicable here.

PAIN IN THE ACHILLES TENDON AREA

Evaluation

See Figure 17-9.

PAIN IN THE ACHILLES TENDON AREA

Posterior Calcaneal Bursitis (Pump Bump) *(Uncommon)*

Diagnosis

A tender bump is found over the achilles tendon (Figure 17-9).

Pathophysiology

A bursa forms and becomes irritated from the upper border of a heel counter that is too firm and/or convex.

Treatment

The *offending shoe* must be *avoided*, with or without the temporary use of shoes with a completely open heel to allow some resolution.

If severe, a small amount of *steroid* may be *injected* into the bursa. Never inject into the achilles tendon itself. See Chapter 3 for further details and precautions.

In chronic cases, referral for *surgical excision* can be offered.

Retrocalcaneal Bursitis *(Rare)*

Diagnosis

Tenderness and sometimes swelling is found anterior to the achilles, behind the talus. A common problem in dancers.

Pathophysiology

Irritation is usually from a spur off the talus, or can be associated with achilles tendonitis.

Rarely due to an unrecognized fracture of the posterior talus diagnosed as an "ankle sprain." Bone scan may be necessary for diagnosis.

Treatment

Treat as for achilles tendonitis.

Recalcitrant symptoms if associated with a spur or fracture should be referred for surgery.

Achilles Tendonitis *(Common)*

Diagnosis

Pain and tenderness are located at the achilles tendon insertion (Figure 17-9) or along the tendon a few centimeters proximal.

Be sure the tendon itself is intact: no gap should be palpable, and the ankle should plantar flex if the calf is squeezed with the patient prone and relaxed. Otherwise a partial rupture may be present and non-weight bearing and immediate orthopedic consultation are indicated.

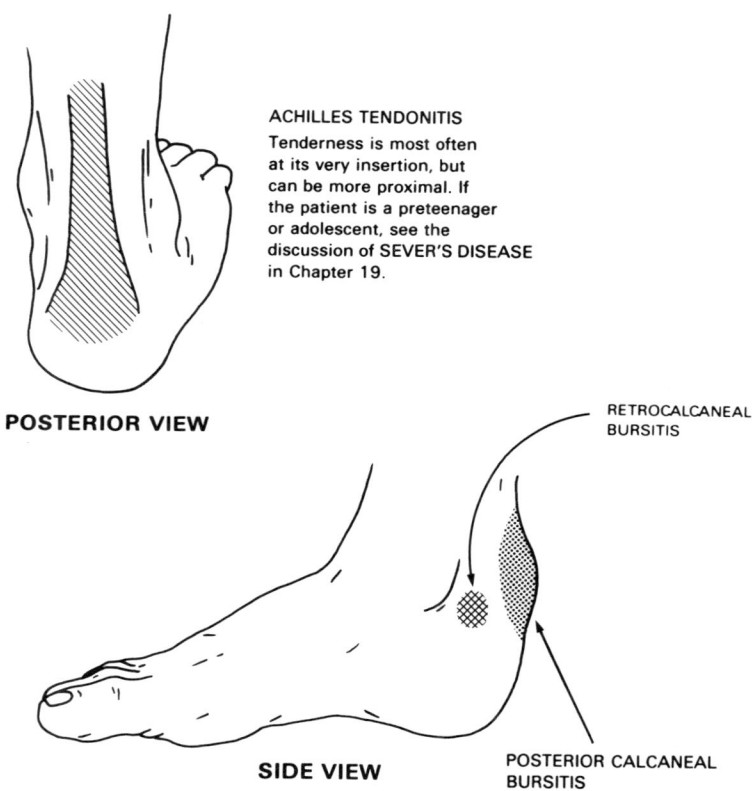

Fig. 17-9. Sites of tenderness: posterior heel.

PAIN IN THE ACHILLES TENDON AREA

Pathophysiology

Excessive stretch occurs from unaccustomed activity, especially with much uphill walking, or from a new shoe with a heel lower than the patient had been wearing. Calf tightness, inflexible shoes, and hyperpronation are contributing factors. Inflammation may be at the insertion or higher up, where it actually is in the "paretenon" that surrounds the tendon.

Treatment

If inflammation and pain are severe, the patient is at risk of his achilles rupturing with continued activity and *rest* should be prescribed. Consider placing the patient in a *short-leg walking cast* with the ankle slightly plantar flexed for a few weeks.

In less severe cases, have the patient *rest* enough to allow matters to start improving.

Uphill and irregular surfaces should be avoided.

Flexible shoes should be worn.

A *calf stretching* program as shown in Patient Handout 20-2 should be begun.

A one-fourth inch *heel lift* should be added to both shoes.

Correct hyperpronation (see box in Chapter 18).

Advise 20 minute *ice* applications, and use *antiinflammatory mediction* as needed.

Steroid injection is contraindicated because it can predispose to rupture.

Achilles Strain *(Uncommon)*

Diagnosis

Following an acute traumatic incident, the patient complains of pain, and tenderness is found in the achilles tendon.

Often associated with ankle sprains.

If any gap is palpable, or ecchymosis is noted, or the foot doesn't passively plantar flex when the calf is squeezed with the patient prone and relaxed, a partial or complete tear may be present which requires immediate orthopedic consultation and probably surgical repair.

Pathophysiology

An "acute overload" injury to the achilles.

Treatment

See the caution above under "diagnosis."

Otherwise, significant injuries are best treated with several weeks in a *cast* with the ankle in slight plantar flexion, followed by rehabilitation supervised by a physical therapist.

Mild injuries should be treated as for achilles tendonitis, above.

PATIENT HANDOUT 17-1

Your Sprained Ankle

If your ankle is sprained it means the ligaments that hold the bones together have been injured. How soon you get back to normal and how much trouble this will cause you later depends of course on how bad the injury, but also very much depends on you: on how well you take care of it.

1. Ice

For the first 48 hours put an *ice pack* on your ankle for twenty minutes at a time, as often as you can: every hour would be okay. Later, use ice applications whenever it is hurting and the swelling seems to be increasing. Never put ice directly on the skin, and never put heat on an injury.

2. Compression

This also helps keep the swelling down, and the less swelling you have the less trouble you'll have. Keep an *elastic bandage* on just about all the time as long as there is swelling.

3. Elevation

Again, to control swelling. Keep your leg higher than your heart as much as you can for the first two days or so; later, so needed.

4. Rest and Protection

Your doctor will have discussed with you whether you need crutches or a cast. Follow his or her instructions as to when you can start putting weight on the ankle. When you start, don't overdo: build up the stress on your leg slowly and easily, resting or going back to the crutches when necessary.

Certainly stay off of uneven ground and high heels. Don't run until you can walk without pain; and don't take part in activities requiring sudden movements and turns until you can run without pain and have tried quick turns and such on your own.

5. **Exercise**

   On the day after your injury, begin moving your ankle straight up and down ten times an hour or so. As soon as the ankle starts to feel better, switch to an exercise where you draw the alphabet, one letter at a time, with your big toe every hour. Draw the letters as big as you can. These exercises will help pump the swelling out and prevent stiffness.

6. **For athletes:** preventing trouble later

   Don't go back to activity until your doctor says okay. When you do:
   - wear an elastic bandage, or much better, tape your ankle. (You will be shown how).
   - choose high-top shoes. If you must wear cleats, they should be as widely placed and numerous as possible.

PATIENT HANDOUT 17-2

If You Keep Spraining Your Ankle

Here are some things that you can do to keep from reinjuring your ankle. All of these may help; none are a cure-all. If problems persist, report to your doctor.

1. Stay off of inclined or hilly or lumpy ground, and certainly off of surfaces with holes.
2. Wear high-top shoes. Certainly stay off of high heels. If you wear cleats, be sure they are as widely spaced and numerous as possible. Your doctor may also recommend wedges.
3. Wear an elastic bandage on your ankle. This will help give you better "joint-feel" and remind you to be careful. But don't be fooled . . . it's not strong enough to prevent your ankle from turning if you step wrong.
4. Taping your ankle before athletic activity is a must. You will be shown how. Commercially available splints and wraps are much more convenient, but may not always be as effective as well-applied tape.
5. Do the following exercises faithfully.

To strengthen your ankle:

A. put elastic around your toes and a table or chair leg, and twist your ankle in the direction of the little toe. Do it slowly and repeat three sets of ten.

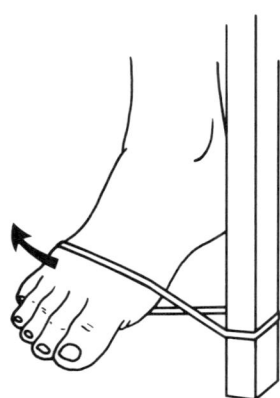

PATIENT HANDOUT

B. Stand on one foot, hold on to something with your hands only for balance. Raise up on your toes slowly, and let back down slowly. Go for three sets of ten.

To stretch your heel cord:

> After you are warmed up, lean forward against the wall with both hands and keep your knees straight and your feet flat on the floor. Lean forward until you feel a tightness (but not quite pain) in your calves. Hold for about a minute, without bouncing. Repeat with your knees bent a bit. (You'll feel this one in your achilles.)

To improve balance and joint feeling:

Stand and balance on one leg at a time. Try to increase the time you can do this to two minutes.

PATIENT HANDOUT 17-3

How to Tape Your Ankle

Use heavy cloth tape, ½-in. wide.

Start on the inside of your leg, above your ankle, and go down under your heel and up the outside of your leg to above your ankle. The trick is to keep your ankle turned outward and maintain tension on the tape througout. Add more strips, each one about half overlapping the one in front. Then anchor these strips with some overlapping strips going ¾ of the way around the leg.

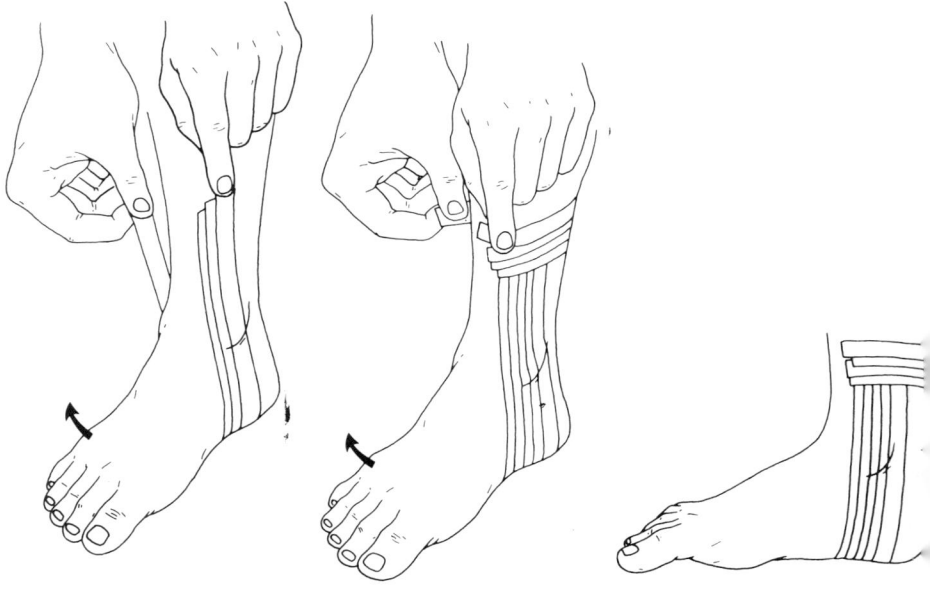

18

The Foot

History	230
Examination	230
Diffuse Foot Pain	231
Evaluation	231
Acute Foot Strain	232
The Biomechanics of Walking	232
Chronic Foot Strain in the Hyperpronating Foot	233
Chronic Foot Strain in the Rigid, High Arched Foot	235
Diffuse Degenerative Arthritis of the Foot	235
Controlling Pronation	236
Inferior Heel Pain	236
Evaluation	236
Plantar Fascitis	236
Painful Heel Pad	238
Pain in and Around the Ball of the Foot	239
Evaluation	239
Hallux Valgus and Bunion	239
Degenerative Arthritis of the First MTP Joint	241
Sesamoiditis	242
Transverse Arch Pain	242
Evaluation	242
Metatarsalgia	244
Morton's Neuroma	244
Dorsal Foot Pain	245
Evaluation	245
Stress Fracture of a Metatarsal Bone	245
Extensor Tendonitis	246
Toe Problems	246

18. THE FOOT

Hammertoes	246
Ingrown Toenail	247
Keratoses	248
Evaluation	248
Callus	248
Soft Corn	248
Hard Corn	248
Plantar Wart	250
Foot Lumps	250
Foot and Toe Trauma	250
Foot Sprain	251
Fracture of the Base of the Fifth Metatarsal	251
Toe Fracture	252
Sesamoid Fracture	252

Foot and heel problems in *infants* and *children* are covered in Chapter 19. Foot problems in *runners* are specifically discussed in Chapter 20. *Numbness and weakness* of the foot are covered in Chapter 13. *Foot cramps* are covered in Chapter 21. Problems around the *ankle* are the subject of Chapter 17. *Contracture* of the plantar fascia is analogous to Dupuytren's contracture in the hand. See Chapter 9.

For the anatomy of the foot see Figure 18-1.

HISTORY

Ask about the *duration* of symptoms.
Has there been any *specific trauma, strain,* or *overuse?*
What types of *shoes* are worn? Has there been a recent change?
Have the patient show you the *location* of the pain.
Is there any *paresthesia* or *weakness?*
Inquire about *precipitating* and *relieving activities* and types of *footwear.*
Is there any previous history of gout or other joint problems?
Is the patient *diabetic?*

EXAMINATION

Look for specific sites of *tenderness,* and if any are found refer to the appropriate section in this chapter or the chapter on the ankle.

Test *range of motion* of the various joints.

Observe the patient while he or she is standing or walking to detect *loss of longitudinal arch* and *excessive pronation.*

Look at the patient's *shoe,* especially noting localized wear, poor fit, or out-of-the-ordinary shape.

Check *neurovascular status.*

DIFFUSE FOOT PAIN

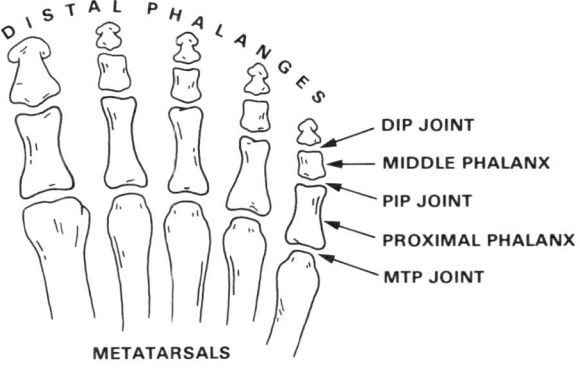

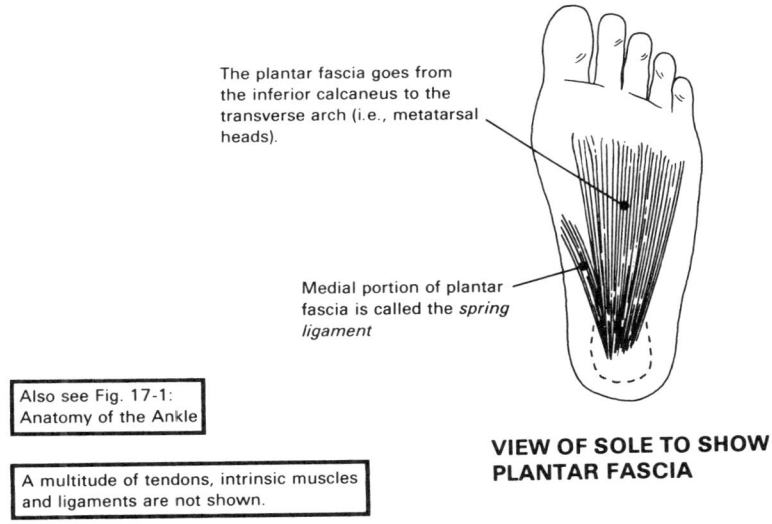

Fig. 18-1. Anatomy of the foot.

DIFFUSE FOOT PAIN

Evaluation

See the recommended history and examination at the beginning of this chapter.

Note that poorly localized foot pain with or without neurologic symptoms

18. THE FOOT

may be due to vascular problems, peripheral neuropathy or any of the conditions covered in Chapter 13. Specifically, diffuse pain in the heel or sole can be due to the TARSAL TUNNEL SYNDROME.

It goes without saying that chronic foot pain can be due to ill-fitting shoes, and this possibility should be looked for first. Otherwise, chronic or persistent diffuse foot pain is more often found in the hyperflexible (hyperpronating) foot or the especially rigid (high-arched) foot than in feet with average flexibility. Either of these two problems of course can also be major contributing factors to localized pain at various locations in the foot and leg.

**The Biomechanics of Walking
(or: pronation and all that)**

In order to understand the overuse and degenerative conditions that afflict the human foot and leg, one must understand how they function in their basic job, walking. Also, the best treatment of many conditions of lower extremity includes biomechanical corrections to alleviate excessive stress on painful structures. Therefore, the following very simplified explanation:

Imagine a person walking. Picture him or her at the moment when one leg is as far forward of the other leg as it gets, and lets "walk through" a "cycle of gait" from there.

Lets assume the right leg is the trailing one. The next motion necessary is obviously for the right leg to swing forward to touch the ground in its turn. (Meanwhile, of course, the left leg is in contact with the ground, pushing backward and propelling our subject in the direction he or she wants to go.) Now when the right leg is trailing behind, the pelvis is obviously turned too, with the right pelvis further behind; so when the right leg swings forward, it must *internally rotate* (think about it). (Meanwhile, the hip is flexing and the knee is extending and the ankle is dorsiflexing.)

Anyway, when the right heel hits the ground the right leg is still in the process of internally rotating. So the last bit of internal rotation of the leg, as it gradually and smoothly decelerates that motion, occurs with the heel fixed to the ground and unable to turn. So what happens is that the internal rotation of the tibia is converted into eversion of the calcaneus (tilting bottom out) through the subtalar joint, which is constructed at an angle to make this possible.

Now when the calcaneus everts, other things happen: the midtarsal joints unlock, which in turn allows the arch to flatten. This entire process of internal rotation of the leg, eversion of the heel, unlocking of the midtarsal joints and flattening of the arch is what is called *pronation* of the foot. Pronation allows the foot to conform to the ground, and also helps absorb some of the shock.

How much pronation occurs will depend on several anatomic factors: the angle of the subtalar joint, how tight are the ligaments that hold the foot together, and the alignment of the leg. It will also obviously be affected by the terrain that the foot is stepping on.

People who pronate less than average have higher arched, more rigid feet which don't absorb shock as well. People whose feet pronate more than average have more

DIFFUSE FOOT PAIN

> motion of, and stress on, various structures, such as the metatarsals, plantar fascia, achilles tendon, and posterior tibialis muscle.
>
> Keep in mind that there is no strict dividing line between people who pronate and those who don't; it is a continuum. Women in general tend to pronate more than men.
>
> Also note that an occasional person will have a rigid flat foot, which actually falls more into the category of congenital or developmental defects. And some feet are constructed so that the forefoot may be in more varus or valgus alignment than the rear foot.
>
> Back to our subject. He now wants to push off with his right foot. To do that efficiently the foot must be rigid; and we have already seen that a pronated foot is flexible. So the foot moves back into *supination* before pushoff; the calcaneus inverts (tilts bottom-in), the midtarsal joints lock and the arch becomes rigid. The large toe dorsiflexes and provides some of the power of pushoff. (Meanwhile the hip is extending, the knee flexing and the ankle plantar flexing.)
>
> Observing a patient stand and walk can give a good idea of how much he or she pronates, and examining the shoe wear pattern can be very helpful too. See Figures 18-2 and 20-1.

Acute Foot Strain *(Common)*

Diagnosis

Diffuse foot pain follows unaccustomed standing or walking.
Examination is normal except possibly for mild tenderness in various locations.

Treatment

A period of *rest* and *elevation* is usually all that is needed.
Ice packs or *antiinflammatories* may be helpful in more serious cases.

Chronic Foot Strain in the Hyperpronating Foot
(flat feet, pes planus) *(Very common)*

Diagnosis

The patient may complain of diffuse foot pain, or a variety of more specific symptoms may coexist: Morton's Neuroma, bunion, plantar fasciitis, posterior tibialis tendonitis, achilles tendonitis, medial distal shin pain, or patellofemoral pain.

On examination, the patient will be found to have a flexible hyperpronating foot (see Figure 18-2). Examination of the patient's shoes should confirm excessive pronation (see Figure 20-1).

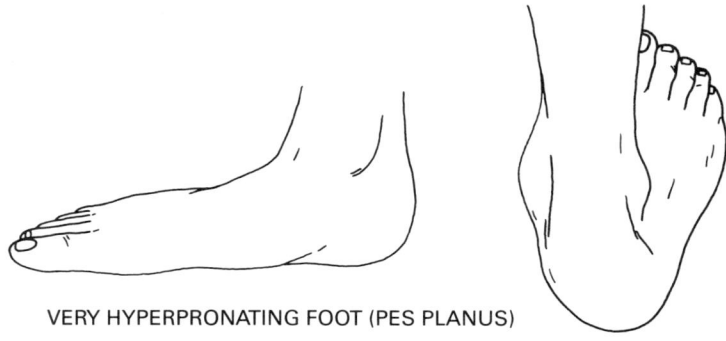

VERY HYPERPRONATING FOOT (PES PLANUS)

Note: – calcaneal eversion
(Achilles points
bottom out.)
– loss of longitudinal arch
– loss of transverse arch
– valgus of forefoot

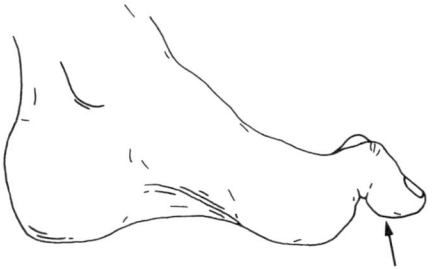

Hammertoes may be present.

VERY HIGH-ARCHED RIGID FOOT (PES CAVUS)

Fig. 18-2. Extremes of foot architecture.

Pathophysiology

The excessive motion in this foot causes extra stress on, and inflammation of, various structures.

Treatment

See the box below on "controlling hyperpronation."

Symptomatic treatment including rest, ice, warm massaging baths and antiinflammatories may also be offered.

DIFFUSE FOOT PAIN

Chronic Foot Strain in the Rigid, High-Arched Foot
(pes cavus) *(Common)*

Diagnosis

The patient complains of diffuse foot pain and possibly localization to the metatarsal heads, inferior heel or other locations.

Examination reveals a more or less high-arched rigid foot (Figure 18-2).

Pathophysiology

These feet do not absorb shock well, tend to concentrate stress in various locations and send more of it up the leg, and are prone to premature degenerative change.

Treatment

Shoes with good shock absorption characteristics should be chosen, and must be well-fitted. Consider adding a full-length visco elastic pad.

Symptomatic treatment such as aspirin may also be recommended, as well as weight loss if indicated.

Chronic Foot Strain Due to Other Deformities
(Uncommon)

This includes conditions such as the congenitally rigid flat foot, tarsal coalition (a bony or cartilagenous bridge between the calcaneus and the talus or navicular, leading to restricted movement), forefoot valgus or forefoot varus. These problems are best left to the podiatrist.

Diffuse Degenerative Arthritis of the Foot *(Uncommon)*

Diagnosis

Diffuse pain is worse after weight bearing.
X-ray will reveal degenerative change.

Pathophysiology

See Chapter 22. Degenerative arthritis is accelerated in the rigid foot and the hyperflexible foot.

Treatment

Advise weight loss if the patient is obese.
Use rest, ice, and antiinflammatory medications as needed.
Treat hyperflexibility or rigidity as indicated above.
Steroid injection into specific troublesome joints may be offered.

> **Controlling Pronation**
>
> If a patient has a condition known to be exacerbated by excessive pronation, and the patient does indeed pronate more than average, controlling the excess foot movement can be quite helpful. This can be accomplished by:
>
> - selecting shoes with the following characteristics:
>   1. snug, stiff heel counter
>   2. built on a curved last
>   3. good arch support
> - adding medial heel wedges (Figure 18-3) or OTC arch supports
> - best of all, by the molding of customized orthotics after taking a plaster impression and then adding appropriate correction. This is best done by a podiatrist, and can be expensive.
>
> Patients with other biomechanical abnormalities such as forefoot varus or valgus can also sometimes benefit from orthotics.

INFERIOR HEEL PAIN

Evaluation

See Figure 18-4.

Lancinating type pain, along with tenderness more medial than usual, suggests entrapment of the medial calcaneal nerve, which may be difficult to treat conservatively.

If no tenderness is found the pain may be referred from DEGENERATIVE ARTHRITIS OF THE SUBTALAR JOINT, covered in Chapter 17.

Acute inflammation in the heel area can be a manifestation of REITER'S SYNDROME (Chapter 23), or rarely, of GOUT (Chapter 24).

For problems around the ankle and posterior heel, see Chapter 17.

Plantar Fasciitis (Heel Spur) *(Common)*

Diagnosis

The patient complains of inferior heel pain often with radiation forward into the sole or arch and worse with prolonged weight-bearing.

Commonly there will be a good deal of pain for the first few steps after a period of non-weight-bearing (upon rising in the morning, for example).

On examination, there is tenderness anywhere along the plantar fascia insertion (Figure 18-4).

INFERIOR HEEL PAIN

DONUT PAD
Available in many shapes and sizes, or can be cut out to fit. The hole goes over the painful spot; the donut around it.

An example of ACCOMMODATIVE PADDING. A pad with areas cut out to accommodate prominences and distribute pressure around a painful area rather than on it.

MEDIAL HEEL WEDGE
Tapered, thicker at edge. Helps decrease pronation. May predispose to ankle sprains.

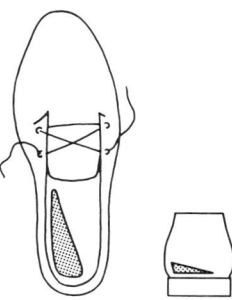

LATERAL HEEL WEDGE
Useful in recurrent ankle sprains. Increases pronation.

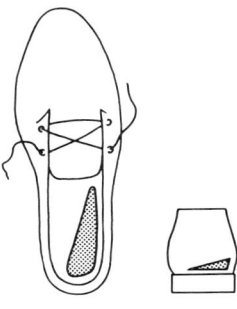

Fig. 18-3. Therapeutic devices for the foot.

Often pain can be increased by forcibly dorsiflexing the toes and forefoot, thus stretching the plantar fascia.

X-ray may or may not show a spur. The spur, if present, is evidence of long-term excessive strain on the insertion of the plantar fascia, but it is not in itself the etiology of the patient's pain. (It points forward not downward).

Look for evidence of hyperpronation, and check passive ankle dorsiflexion with the knee flexed (approximately 20° is considered adequate).

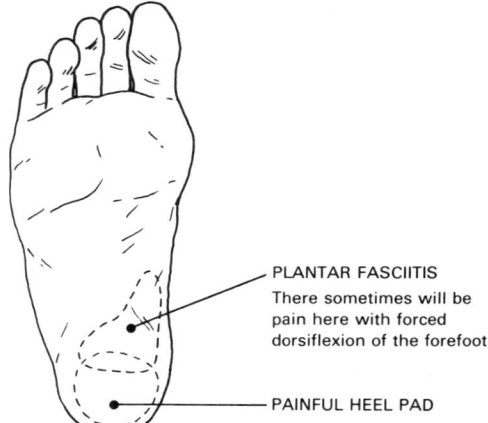

Fig. 18-4. Sites of tenderness: inferior heel. (Tenderness may be anywhere within the areas shown.)

Pathophysiology

Inflammation occurs where the plantar fascia inserts into the calcaneus due to prolonged microtrauma. More common in the hyperpronating or cavus foot.

More often seen in older or obese patients, athletes and dancers.

Treatment

A *visco elastic heel pad* should be inserted into the shoe.

Calf and *heel cord stretching* is beneficial. (Patient Handout 20-2).

Correct any *hyperpronation* (see box above). In the cavus foot, an arch support can be helpful.

Advise 20 minute *ice* applications regularly and as needed.

Prescribe *aspirin* or an *NSAI*.

In prolonged cases consider *steroid injection* (Figure 18-5) or a course of *ultrasound*. Both have only fair success rates.

Advise *weight loss* for the obese.

The natural course is usually of spontaneous resolution, though sometimes only after many months. Referral for *surgical release* should be considered beyond that time.

Painful Heel Pad (Stone Bruise) *(Uncommon)*

Diagnosis

Pain and tenderness occur diffusely under the entire heel or the posterior part of it.

PAIN IN AND AROUND THE BALL OF THE FOOT

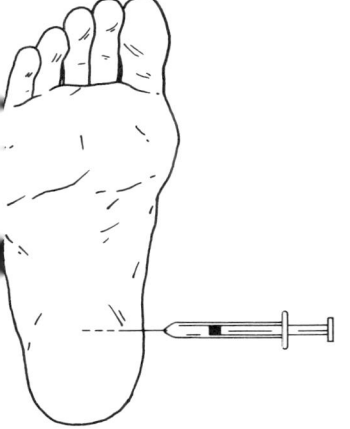

Fig. 18-5. Injection for plantar fasciitis. After prepping the area and spraying with ethyl chloride, inject toward the point of maximal tenderness from the medial approach. (This is much less painful than the inferior approach.) Use a long 25-gauge needle and inject about a cc of steroid with a cc of local anesthetic. See pp. 24-26 for further details and precautions.

Pathophysiology

The heel pad loses its elasticity as the body ages, and acute or repetitive strain or trauma can cause the entire layer of fatty and fibrous elastic tissue to become inflamed.

Treatment

Place a viscoelastic heel pad in the shoe.
Oral *antiinflammatory medications* or *injection of steroid* may be tried.

PAIN IN AND AROUND THE BALL OF THE FOOT

Evaluation

Sudden atraumatic onset of pain and swelling in the first MTP joint is almost pathognomic of GOUT. See Chapter 24.

Look for a HALLUX VALGUS condition (Figure 18-6) and a tender or nontender thickening on the medial aspect of the joint. This is a BUNION.

Hallux Valgus and Bunion *(Common)*

Diagnosis

HALLUX VALGUS refers to lateral angulation of the great toe at the MTP joint.

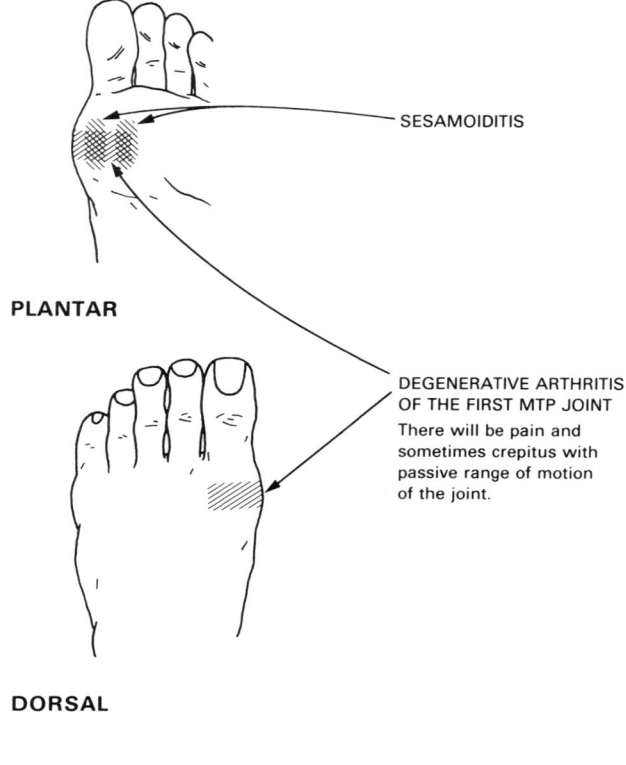

Fig. 18-6. Sites of tenderness: ball of foot.

A BUNION is a chronically thickened and continually or intermittently inflamed bursa on the medial aspect of the joint.

The analogous condition occurring at the fifth MTP joint is known as a Taylor's bunion or bunionette.

Pathophysiology

Although some congenital malformations (metatarsus primus varus or a deformed first metatarsal head) predispose to hallux valgus, the usual direct cause is the wearing of shoes with narrow toe areas.

Excessive motion of the first "ray" that occurs in the hyperflexible foot may also contribute.

Excessive pressure is then placed on the medial aspect of the first MTP joint, with the resultant development of a thickened, inflamed bursa.

Treatment

The wearing of *shoes with wide toeboxes* should be advised, and a *donut pad* (Figure 18-3) or lamb's wool may be helpful.

In the hyperflexible foot, *correcting* the excessive *pronation* may help to retard progression.

Steroid injection into an inflamed bursa may give some relief.

Associated degenerative arthritis of the joint should be treated as outlined below.

Referral to a podiatrist or orthopedist for consideration of *surgery* is often necessary.

Degenerative Arthritis (Osteoarthritis) of the First MTP Joint *(Common)*

Diagnosis

Variable diffuse joint line tenderness is found, as well as variable pain with range of motion of the joint, often with crepitus (Figure 18-6).

Range of motion usually becomes restricted with time ("hallux rigidus").

The X-ray may be normal early.

A uric acid level should be drawn to help exclude the diagnosis of gout.

Pathophysiology

Accelerated wear and tear to this joint occurs from excessive flexion/extension or is secondary to deformity such as hallux valgus.

It is common in people who wear high heels, in runners, in soccer players, and in dancers.

Treatment

Stiffening the sole (with a stiff insert or with steel-shank shoes) helps by decreasing the motion at the joint.

It is important that *low heels* be worn.

Oral *antiinflammatory medications* or intraarticular *steroid injection* (Figure 18-7) may help the patient over an acute exacerbation.

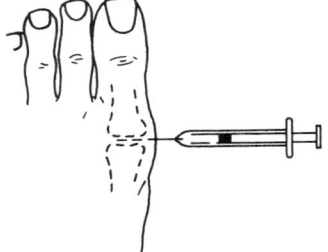

Fig. 18-7. Injection into the first MTP joint. Feel the joint line over the dorsal-medial aspect with your fingertip. Moving the base of the big toe up and down can help localize it. After sterile prep and ethyl chloride spray, enter the joint with a short 25-gauge needle and inject a cc of steroid and a cc of local anesthetic without epinephrine. See pp. 24-26 for further details and precautions.

Referral for *surgery* may be necessary if the measures noted above do not result in sufficient relief.

Sesamoiditis *(Rare)*

Diagnosis

Pain and tenderness are limited to the area under the sesamoid bones (Figure 18-6).

Pain often results if the involved sesamoid is compressed upward while the great toe is passively dorsiflexed.

These bones are often bipartite, and this should not be confused on X-ray with a fracture.

Pathophysiology

Pain is due to excessive pressure or to roughening of the superior surface of the sesamoid, with consequent friction as the toe is flexed and extended.

Treatment

A *stiff sole, low heel,* and *donut pad* (Figure 18-3) are helpful.

Steroid injection with a #25 needle from the medial side may be offered. See Chapter 3 for details and precautions.

Patients who wear cleated shoes should be sure there isn't a cleat directly under the sore area.

TRANSVERSE ARCH PAIN

Evaluation

See Figure 18-8.

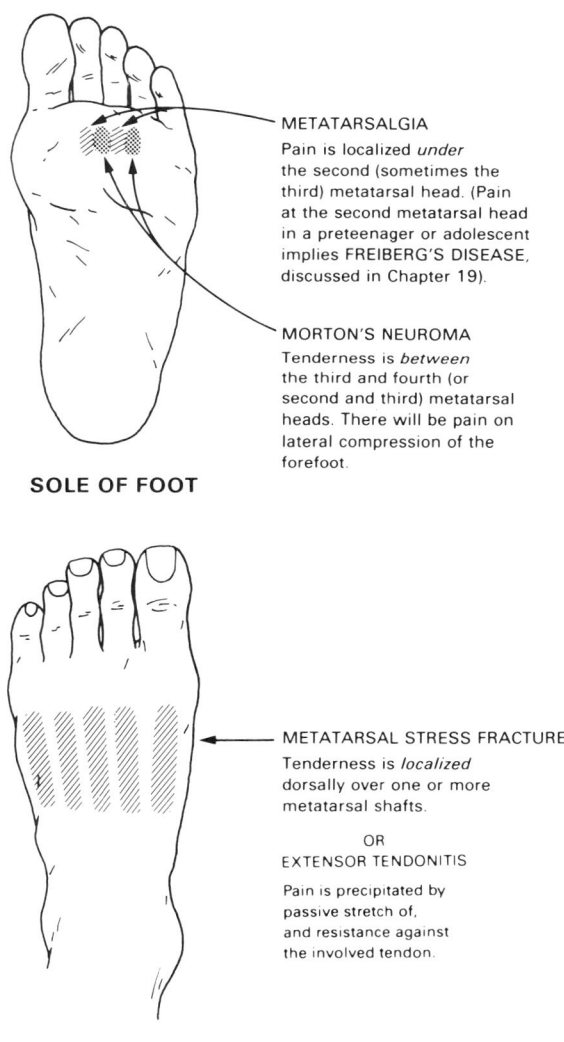

Fig. 18-8. Sites of tenderness: forefoot.

Metatarsalgia *(Common)*

Diagnosis

Tenderness is located under the second (sometimes the third) metatarsal head, often with callus formation as well.

Pathophysiology

Chronic foot strain in the hyperpronating or cavus foot changes weight-bearing in such a way as to place more pressure at this point.

Often a change in shoe (especially to a higher heel) will be the precipitant.

Treatment

A *low heel* should be worn.

A donut pad often helps.

Correct any excessive pronation.

In severe cases a course of an *antiinflammatory medication* or local *steroid injection* should be offered.

Referral to a podiatrist for more precise biomechanical correction should be considered in refractory cases, with surgery as a last resort.

Morton's Neuroma* (Interdigital Neuroma) *(Common)*

Diagnosis

Tenderness is *between* two of the metatarsal heads (Figure 18-8).

Pain is produced by lateral compression of the foot.

Sometimes there is shooting pain or numbness in the interspace distal to the neuroma.

The patient will report that relief is obtained more by removing the shoe than by getting off the foot.

Pathophysiology

This is a fusiform swelling of the interdigital nerve, usually secondary to compression by the metatarsal heads in a hyperpronating foot or an overly tight shoe.

Occasionally direct pressure from below can irritate.

Treatment

Shoes must be *wide* enough so that further irritation of the nerve is avoided.

Any hyperpronation should be corrected.

---

*Not to be confused with MORTON'S FOOT, which is a congenitally short first metatarsal and is named after a different Morton.

DORSAL FOOT PAIN

A donut or accommodative *pad* or visco elastic sole pad may help.
Steroid injection may be offered. Success rate is fair. See Figure 18-9.
Often referral to a podiatrist or orthopedist for *excision* becomes necessary.

DORSAL FOOT PAIN

Evaluation

See Figure 18-8.

Stress Fracture (March Fracture) of the Metatarsal Bone *(Common)*

Diagnosis

Forefoot pain comes on after overuse (e.g., a runner who has increased his mileage or a new soldier forced to take long marches) and decreases somewhat with rest.

On examination, well localized tenderness is found on one or more of the metatarsal shafts on the dorsum of the foot (Figure 18-8).

Pathophysiology

See Chapter 20.

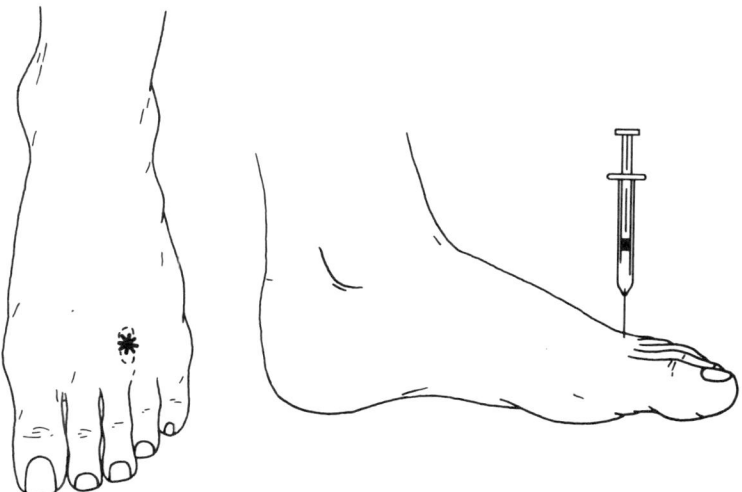

Fig. 18-9. Injection for Morton's neuroma. Localize the neuroma on the *plantar* surface between the metatarsal heads; now place your finger on the *dorsal* surface directly superior to the neuroma. After prep and ethyl chloride spray, inject *directly downward* with a 25-gauge needle about a cc of steroid and a cc of local anesthetic without epinephrine. See pp. 24-26 for further details and precautions.

Treatment

Decrease weight-bearing activity. If stress on the injured bone continues unabated, the lesion may extend and become a real through and through fracture.

A short leg walking *cast* for two to four weeks is unnecessary but may be used for symptomatic relief.

Analgesia should be used as needed.

On gradual return to activity, avoiding hilly ground, using a visco elastic pad and correcting any hyperpronation may help prevent recurrence.

Extensor Tendonitis *(Uncommon)*

Diagnosis

Tenderness is localized over a tendon.

Pain is reproduced with passive flexion of the toes or resisting the patients attempt to extend the involved toes.

Pathophysiology

Often from shoe pressure.

Treatment

Avoid shoe pressure.
Use ice and antiinflammatories as needed.

TOE PROBLEMS

Hammertoes *(Common)*

Diagnosis

By inspection (Figure 18-2).

Pathophysiology

May be secondary to pes cavus or may be due to the long-term wear of shoes and/or socks that are too short and crowd the toes into flexion.

The weakening of the intrinsic flexors that comes with shoe wear may be a contributing factor in those patients whose anatomy predisposes them to the problem.

Treatment

Shoes with a *high toe box* should be obtained.

Treat the complications, metatarsalgia and corns, as discussed under those headings.

Referral for *surgery* may become necessary.

TOE PROBLEMS

Ingrown Toenail *(Very Common)*

Diagnosis

Acute or chronic inflammation occurs at the nail margin.

Pathophysiology

If the corner of the nail is trimmed too aggressively, the nail can grow into the overhanging skin.

A very convex nail predisposes to the problem.

Treatment

In acutely inflamed cases, treat with an *antibiotic* effective against staphylococcus as well as with *elevation, open shoes,* and twice daily *warm soaks*.

If the edge of the nail can be pried out of the skin, daily insertion of a piece of cotton between the nail and the skin should be performed by the patient until the corner of the nail has grown past the overhanging tissue. (This should be done after soaking, and with a blunt instrument such as the edge of a tongue blade.)

If there is a *spur* protruding into the soft tissues (often from previous efforts at partial nail removal), the method shown in Figure 18-10 may be tried to *remove* it.

If it is decided that a strip of nail must be removed, the corresponding area of matrix must be obliterated or the problem will quickly recur. A variety of methods are used; consult a text on outpatient surgery, or refer the patient.

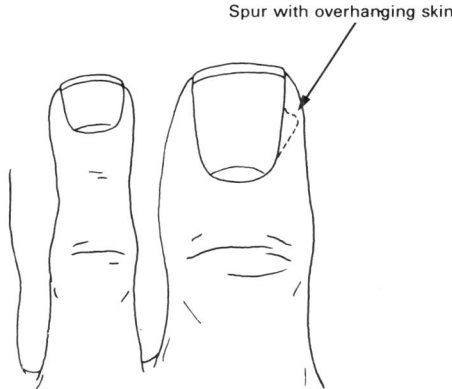

Fig. 18-10. Removing a nail spur. After obtaining informed consent and anesthetizing the toe analogously to Figure 9-9, bluntly dissect the spur out from under the overhanging soft tissue and cut it off using scissors.

To prevent recurrence, patients should be instructed to wear wide enough shoes and always trim their nails straight across (Figure 18-11).

KERATOSES

Evaluation

See Figure 18-12.

Callus *(Very Common)*

Diagnosis/Pathophysiology

This is a diffuse or circumscribed thickening of the skin resulting from excessive pressure or friction.

Treatment

The *underlying cause* of the abnormal pressure should be dealt with.

Shoes should not cause friction at, or put unnecessary pressure on, the callused area.

A *donut pad* (Figure 18-3) should be used.

The callus may be *scraped* (*never* cut into) with a blade or *rubbed* with a pumice stone after *soaking* to soften it.

Soft Corn *(Common)*

Diagnosis/Pathophysiology

These are found between the toes (usually the fourth and fifth). They are caused by pressure from an adjacent toe.

Treatment

As for a callus (above).

Cotton or *lamb's wool* placed between the toes is often helpful.

Forty percent *salicyclic acid plasters* may be applied for 24 hours every few days, always after soaking.

Referral for *surgical excision* of underlying bony prominences may become necessary.

Hard Corn (Neurovascular Corn) *(Common)*

Diagnosis/Pathophysiology

It is sharply demarcated and appears over a bony prominence.

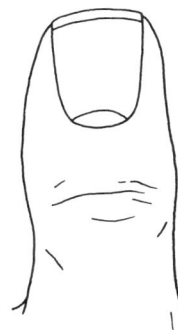

Fig. 18-11. Proper toenail trimming. Note how nail is cut straight across and how corners protrude just a bit instead of digging into the skin.

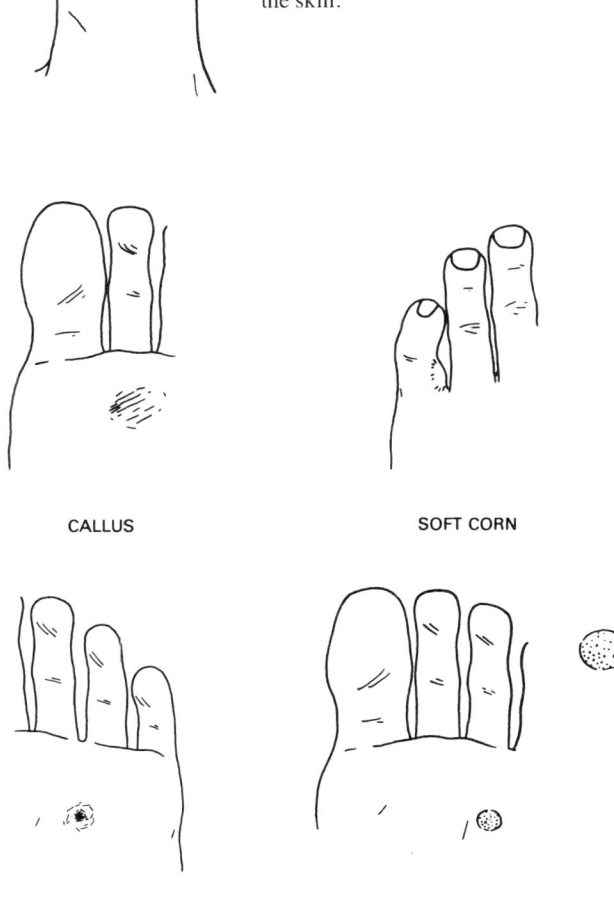

Fig. 18-12. Keratoses.

Treatment

As for a callus or soft corn.

Plantar Wart *(Common)*

Diagnosis

It is sharply demarcated, and not necessarily found over a bony prominence.
Scraping will reveal black dots that are actually capillary endings.
Skin lines go around a wart; they go through a callus or corn.
Warts are painful with lateral compression, while calluses and corns generally are not.

Pathophysiology

Caused by the same papilloma virus that causes warts elsewhere; because of pressure they grow inward rather than outward.

Treatment

Surgery and overly aggressive treatment that will leave a scar should be avoided. (A scar will be permanently painful; at least the wart will eventually go away if the patient holds out long enough.)

Soaking followed by scraping and the application of a keratolytic such as 40% *salicyclic acid plaster*, if repeated every few days, sometimes works.

Liquid nitrogen or *paring followed by the application of trichloroacetic acid* to the base are other methods available.

FOOT LUMPS

GANGLIA are common in the ankle and foot; see the discussion in Chapter 8, which is equally applicable here. Many different small ACCESSORY BONES have been described, and are usually not significant except for the pressure effects they may cause. Diagnosis is by X-ray and treatment by neglect, padding or surgery as necessary. (But see the discussion of the accessory navicular on p. 212.)

FOOT AND TOE TRAUMA

Always examine distal neurovascular status.
Fractures and dislocations not discussed below are considered beyond the scope of this book. Consult an orthopedics text or refer the patient for management.

FOOT AND TOE TRAUMA

Foot Sprain *(Uncommon)*

Diagnosis

The mechanism of injury is usually excessive plantar flexion or dorsiflexion of the toes and forefoot, injuring the dorsal or plantar soft tissues, respectively.
X-ray is normal.

Treatment

Use *elevation* and *ice* acutely.
Advise *compression* by *taping* (Figure 18-13) and limited weight bearing, with resumption of activity as tolerated.
Prolonged pain is usually due to a posttraumatic tendonitis, which may be treated with oral antiinflammatory medications or local injection.

Fracture of the Base of the Fifth Metatarsal
(Jones Fracture) *(Common)*

Fractures of the other metatarsals often require reduction and are not included here.

Diagnosis

The patient reports a sudden inversion stress with the foot in plantar flexion.
Tenderness is found over the base of the fifth metatarsal.
X-ray confirms the diagnosis. Note that a small accessory ossicle is sometimes present at this location, and should not be confused with a fracture. The fracture line is usually perpendicular to the long axis of the metatarsal, the ossicle usually parallel; the line separating the ossicle from the metatarsal never enters the joint at the base of the metatarsal.

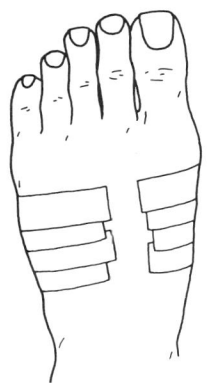

Strips are continuous around bottom of foot.

Fig. 18-13. Taping a foot. Note that the strips of tape about half overlap each other, and that the foot is not circumferentially constricted.

Treatment

Only *symptomatic treatment* is necessary (elevation, ice and compression acutely, and usually a short-leg walking cast applied once the swelling has subsided, left on for 3 to 6 weeks).

However, cases with gross displacement or a large fragment are best evaluated by an orthopedist.

Persistent pain is usually secondary to a posttraumatic peroneal tendonitis, discussed in Chapter 17.

Fracture of the Toes *(Common)*

Diagnosis

The history is one of stubbing the toe or dropping an object onto it.

Treatment

Fractures of the *proximal phalanx* of the great toe or fractures of the distal phalanx that *extend into the interphalangeal joint* should be *referred* to an orthopedist for management.

Fractures of the tuft of the great toe and fractures of the other toes without significant deformity are treated *symptomatically* by *taping* them to an adjacent toe (Figure 18-14) and the use of a roomy or open shoe. A stiff sole is helpful in fractures of the great toe.

Fractures of the Sesamoid of the Great Toe *(Rare)*

Diagnosis

Usually there is a history of a forceful landing on the ball of the foot.

This must be differentiated on X-ray from the smooth edges of a bipartite sesamoid.

Treatment

Use *ice, compression,* and *elevation* acutely.
Then advise *limited weight-bearing* with progression as tolerated.
A donut pad (Figure 18-3) may be helpful.
If pain persists, referral for excision may become necessary.

Dislocation of the IP Joint of a Small Toe *(Rare)*

Diagnosis

Usually there is dorsal dislocation of the middle phalanx on the proximal phalanx.

FOOT AND TOE TRAUMA

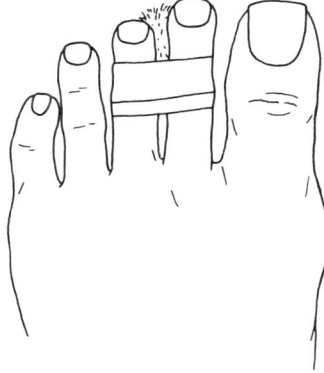

Fig. 18-14. Taping toes. Tape to the adjacent toe (toward the middle toe). Put some lamb's wool or cotton between the toes.

Treatment

Reduction is usually easy without anesthesia (Figure 18-15).
A roomy or open shoe and taping (Figure 18-14) are used symptomatically.

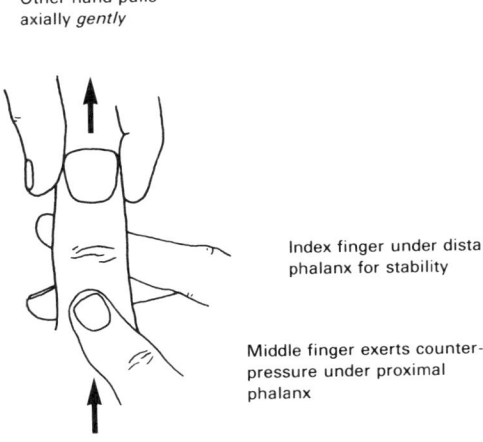

Fig. 18-15. Reducing a dislocated toe.

PATIENT HANDOUT 18-1

What to Look for in a Shoe

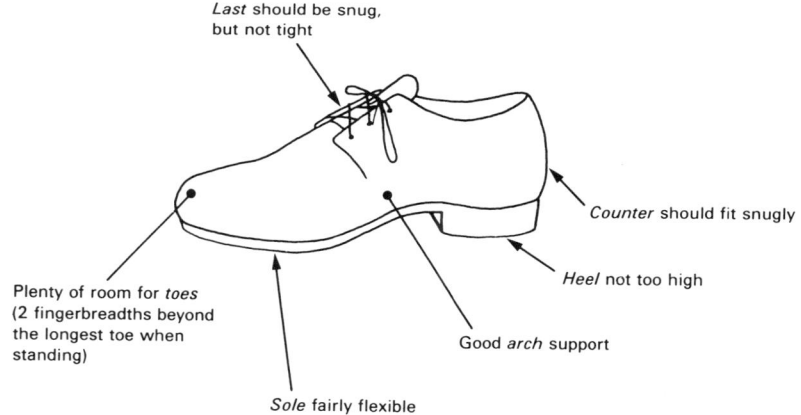

Last should be snug, but not tight

Counter should fit snugly

Heel not too high

Good *arch* support

Sole fairly flexible

Plenty of room for *toes* (2 fingerbreadths beyond the longest toe when standing)

PATIENT HANDOUT 18-2

Taking Care of Calluses and Corns

Calluses and corns usually develop at sites of excessive pressure, either from shoes or from one toe against another. So to prevent them from occurring or worsening:

Wear shoes that don't put pressure on the sore area.

Use a donut-shaped pad to take pressure off the sore are (*Note:* the hole goes *over* the corn or callus, the donut *around* it).

If the problem is between the toes, use a moleskin pad or some lamb's wool to avoid the pressure.

If these measures fail to relieve the problem, and the callus or corn is *painful,* you can do the following:

Soak your foot in warm water.

Then, for calluses, *scrape* it down with a clean blade (never, never cut) or with a pumice stone (available in drug stores).

For corns, or for calluses that don't respond to scarping, get a keratolytic agent at the drugstore (something with *salicyclic acid* in it) and use it as directed.

19

Pediatric Problems

Congenital Foot Problems 258
    Examination 258
    Metatarsus Adductus 259
    Metatarsus Primus Varus 260
Congenital Hip Dysplasia 260
    Evaluation 260
    Congenital Subluxability of the Hip 262
The Pronated Foot in Infants and Children 262
    Evaluation 262
    Flexible Pronated Foot 263
Foot Pain in Children and Adolescents 264
    Evaluation 264
    Freiberg's Disease 264
    Kohler's Disease 264
Heel Cord Pain in Children 265
    Sever's Disease 265
Intoeing Gait 265
    Evaluation 265
    Internal Tibial Torsion 267
    Excessive Femoral Anteversion 267
Bowleg, Knockknee, and Backknee 267
    Evaluation and Management 267
Knee Pain in Children and Adolescents 268
    Evaluation 268
    Osgood-Schlatter's Disease 269
Hip Pain in Children and Adolescents 269
    Evaluation 269
    Transient Synovitis of the Hip 269

CONGENITAL FOOT PROBLEMS

    Legg-Perthe's Disease..271
    Slipped Capital Femoral Epiphysis..................................271
Back Pain in Children and Adolescents 272
    Evaluation..272
    Scheuermann's Disease..272
Abnormal Curvature of the Back 273
    Evaluation..273
    Structural Scoliosis 273
Wryneck in Infants and Children 275
    Evaluation and Management......................................275
Elbow Problems in Children..275
    Radial Head Subluxation..276
    Overuse Injuries Related to Throwing 276
Musculoskeletal Trauma in Children 277
    Torus Fractures of the Distal Radius...............................278
Systemic Illnesses 278
    Juvenile Rheumatoid Arthritis 278
    Rheumatic Fever 279

Musculoskeletal problems in the pediatric age group, aside from trauma, are primarily developmental in nature. Though many of these conditions are self-correcting and require only reassurance, other have great potential for causing lifelong disability if not dealt with aggressively and properly. Only rarely are the young, flexible tissues of the child's musculoskeletal system affected by the degenerative and overuse syndromes that so plague the adult population.

Complex congenital syndromes and bone disease are not dealt with here; the reader is referred to a standard pediatric text for a discussion of these conditions.

Several basic points must never be forgotten when dealing with musculoskeletal complaints in children:

1. Bony tenderness or a complaint of joint or bone pain in a child should be considered to be a sign of *infection, neoplasm,* or *serious occult trauma until proven otherwise.* An X-ray must be performed, and sometimes a bone scan as well.
2. A toxic infant with a fever without obvious cause should always bring to mind the possibility of a *septic joint* (especially hip).
3. A *hemarthrosis* is a possible cause of joint pain and swelling in children that must be considered because of its grave implication of underlying *blood disease or serious joint trauma.*
4. Injuries involving indirect forces will usually occur at the weakest link in the musculoskeletal unit. In the child this is the epiphyseal plate, i.e., the growth plate of bone itself.* Therefore, never make the diagnosis of a sprain, a strain,

---

*As the child grows into adolescence, the muscle belly and musculotendonous junction become the most vulnerable sites; and in adulthood, it is the tendons themselves or their insertions into the periosteum that absorb the brunt of most indirect injury.

or tendonitis in a child without carefully ruling out an *epiphyseal plate injury,* which may not be visible on X-ray and may manifest itself only as point tenderness at the growth plate. These injuries, if not treated properly, can lead to serious problems indeed. (See Table 19-1.)
5. For the same reason, *comparison X-ray views of the uninjured limb* are essential.
6. Multiple injuries or those which do not quite fit with the history given must always bring to mind the possibility of a *battered child.*

One more thing: by far the most important intervention most primary care providers will ever make in caring for the musculoskeletal system (not to mention the rest of) a child is ensuring the use of proper restraints (infant seats, car seats, seat and shoulder belts) while riding in automobiles.

CONGENITAL FOOT PROBLEMS

Examination

If all the borders of the foot are straight, the foot is normal. (Note that the plantar fat pad makes the newborn's foot seem flat, but if it is not pronated (Figure 19-1a), it is normal.)

If there is adduction of the forefoot (Figure 19-1b), see if the deformity is passively overcorrectable. If so, see METATARSUS ADDUCTUS, below. If passive overcorrection is not possible, the infant should be referred to the orthopedist.

If there is adduction of only the first metatarsal (Figure 19-1c), see METATARSUS PRIMUS VARUS, below.

If the foot is in equinovarus (Figure 19-1d) or calcaneovalgus (Figure 19-1e), see if the deformity can be passively *over*corrected. If so, no treatment is necessary, but the flexibility of the foot should be rechecked periodically until the deformity corrects. If the deformity cannot be passively *over*corrected, the child has a CLUBFOOT and must be referred to a specialist immediately, as any delay in proper treatment worsens the prognosis.

If the foot is pronated (Figure 19-1a) and cannot be overcorrected by passive

Table 19-1. Salter's Classification of Epiphyseal Injury

Salter type	Injury
I	Transverse fracture of only the growth plate
II	Fracture through the growth plate and into the metaphysis
III	Fracture through the growth plate and through the epiphysis
IV	Fracture through the epiphysis, growth plate and metaphysis
V	Impaction

CONGENITAL FOOT PROBLEMS

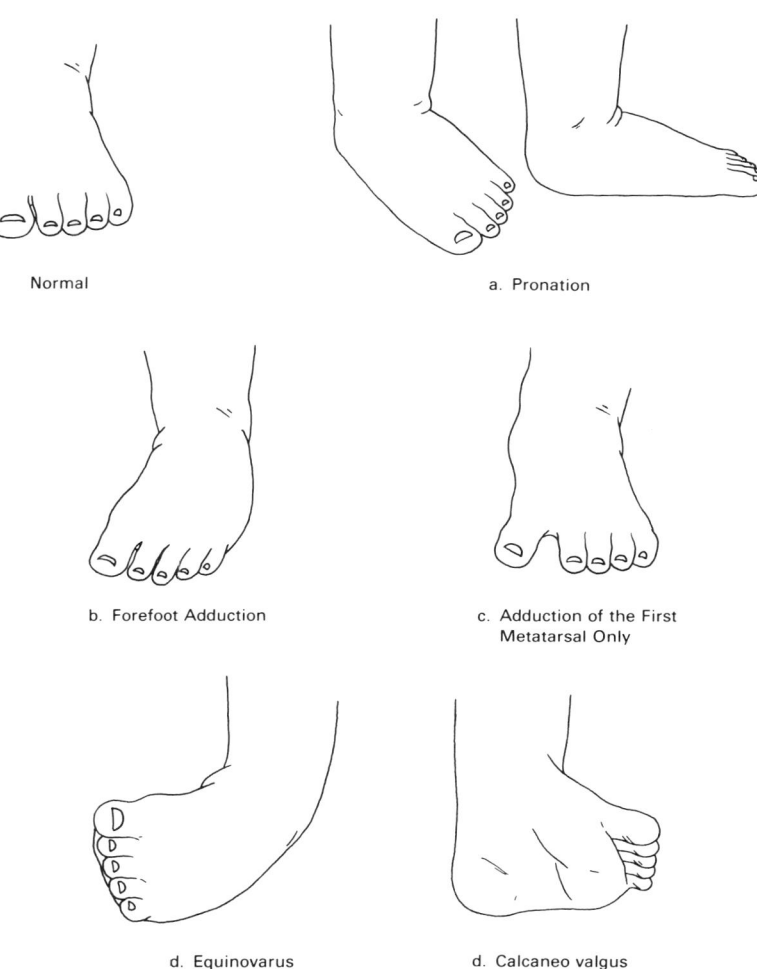

Fig. 19-1. Foot deformities in infants.

manipulation, the patient should be managed by an orthopedist. (Also see the discussion of foot pronation starting on p. 268.)

Metatarsus Adductus *(Common)*

Diagnosis

The lateral border of the foot is convex and the medial border concave, with the head of the talus palpable on the medial side (Figure 19-1b).

The forefoot will flare laterally when the infant's sole is stimulated, and the deformity is passively overcorrectable. (If not, the patient has a form of clubfoot and should be managed by the specialist.)

Pathophysiology

The condition may be secondary to *in utero* positioning.

Treatment

If the abnormality is very mild, no treatment is necessary.

In more obvious deformities, some experts advocate the use of *reverse shoes* (wearing the right shoe on the left foot and vice versa) and *passive stretching* of the forefoot in a valgus direction by the parent several times a day (see Patient Handout 19-1). However, there is a growing belief that no treatment is needed even in these cases, (To reiterate, if the forefoot cannot be passively overcorrected, treatment *is* required and referral to an orthopedist recommended.)

Metatarsus Primus Varus *(Uncommon)*

Diagnosis

The lateral border of the foot is straight, but the first metatarsal points inward, and there is increased space between the first and second metatarsals (Figure 19-1c).

Treatment

None is necessary. However, the parent should be advised never to allow the child to wear shoes with narrow toeboxes, since this will accentuate the deformity and hasten the development of a bunion later in life.

CONGENITAL HIP DYSPLASIA

Evaluation

Ortolani's test (Figure 19-2) should be performed on all infants at birth, and must be repeated as six weeks and three months of age. If the test is positive in the newborn, but abduction of the hip to 90° is possible and there is no other evidence of fixed dislocation (see the following paragraphs), see CONGENITAL SUBLUXABILITY OF THE HIP, below. If the test is positive three weeks or more after birth, the child should be referred to the specialist for management.

Check for a full 90° of passive abduction of each hip in the flexed position (Figure 19-2). This examination should be done both in the immediate postnatal

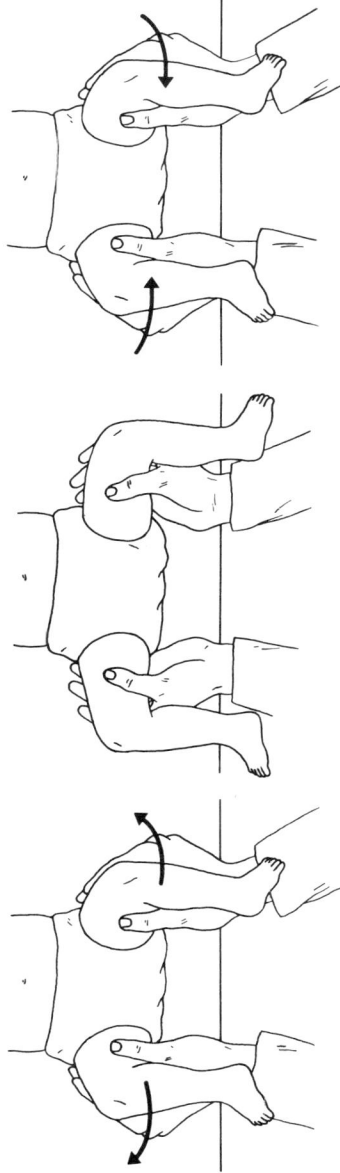

Fig. 19-2. Ortolani's Test and passive hip abduction. Hold one leg in each hand, with both hips and both knees flexed to 90°. Now passively abduct both hips to 90° and return to straight up. An audible or palpable click indicates subluxability of the hip. Lack of a full 90° of passive abduction at any time in the first year mandates referral to an orthopedist.

period and at every well-child visit in the first year. If at any time full passive abduction is lacking, the patient should be referred to the orthopedist.

If the Ortolani maneuver is positive, check for shortening on the affected side (by flexing the hips and noting a discrepancy in the level of the flexed knee), for flattening of the involved buttock with the child prone, or for telescoping of the hip (i.e., the ability of the examiner to move the extended leg cranially and caudally relative to the trunk). These findings are quite rare, but if present imply a *dislocated* hip and mandate immediate orthopedic referral.

Congenital Subluxability of the Hip *(Uncommon)*

Diagnosis

A click is found on performing on Ortolani maneuver (Figure 19-2) in the newborn.

There is full passive abduction of the hip, and no shortening, telescoping, or flattening of the buttock.

Pathophysiology

Intrauterine malpositioning probably plays a role in some cases by limiting the normal motion necessary for proper joint development.

Genetic factors are probably operative as well, as girls with the condition outnumber boys by four to one, and there are different concordance rates in monozygotic and dizygotic twins.

Maternal hormones may have an effect on joint development.

Treatment

The newborn with a congenitally subluxable hip may be treated with *triple diapering* to maintain a frog-leg position, (The idea is to keep the femoral head firmly seated in the acetabulum until the lax ligaments tighten up a bit.)

Recheck the baby at three weeks of age; if the click persists the infant should be referred to the specialist for further treatment.

If there is limitation of abduction or a sign of fixed dislocation at any time, *referral* is mandatory.

THE PRONATED FOOT IN INFANTS AND CHILDREN

Evaluation

The normal infant foot looks flat because of the presence of a fat pad under the longitudinal arch. If examination reveals a flexible foot, this appearance is of no significance.

THE PRONATED FOOT IN INFANTS AND CHILDREN

If in the newborn infant there is limitation of passive plantar flexion and inversion of the forefoot, there will often be a bony prominence on the medial aspect of the foot. This may be due to any of a number of underlying congenital malformations, and these infants should be referred to the specialist for management.

If pronaton of the foot (Figure 19-1a) is noticed when the child begins to bear weight, check the following:

1. Is the achilles tendon too short? Check by seeing if the foot can be passively dorsiflexed past neutral. If not, the foot must pronate just to allow the heel to touch the ground during walking. These patients should be referred to the orthopedist.
2. Does the child have an intoeing gait which he is trying to compensate for? If so, he may be consciously forcing his foot into pronation. Check to see if the foot assumes the normal posture when not bearing weight and whether there is good flexibilty with passive movement. If so, no treatment is necessary except education.
3. If the pronation deformity does not disappear when not bearing weight and cannot be passively overcorrected, the child has a RIGID PRONATED FOOT, which should be managed by the specialist.
4. If the pronation disappears when the child is not bearing weight and can be overcorrected by passive manipulation, the child has a FLEXIBLE PRONATED FOOT.

Flexible Pronated (Flat) Foot *(Common)*

Diagnosis

Pronation can be overcorrected passively by the examiner and disappears when not bearing weight.

A short heel cord is ruled out by normal passive dorsiflexion of the ankle.

Pathophysiology

In the first year or so of weight-bearing, pronation may simply be secondary to the wide-based gait necessary for balance.

If flexible pronation persists beyond that age, it is probably a consequence of ligamentous laxity.

Treatment

Before three years of age, no treatment for the flexible flat foot is necessary. Beyond that age, if the deformity is mild, no treatment is required either.

If the deformity is fairly severe, or if the child complains of foot pain after activity (and no other etiology is found; see the next section), consider referral to a podiatrist for evaluation and orthotic correction. (Again, be sure you are not dealing with a rigid flat foot.)

FOOT PAIN IN CHILDREN AND ADOLESCENTS

Evaluation

Be sure that the problem is not simply poorly fitting shoes creating areas of excessive pressure.

As in any musculoskeletal pain syndrome in children, X-rays must be performed to rule out infection, neoplasm, or an occult fracture.

If the child complains of diffuse pain that is worse after activity, and examination reveals excessive pronation but a flexible foot, see the discussion of the treatment of the mobile pronated foot, above.

Pain and tenderness in the area of the second metatarsal head in a preteenager or adolescent is likely due to FREIBERG'S DISEASE.

Pain and tenderness in the medial foot in a prepubescent or pubescent child suggests KOHLER'S DISEASE.

Freiberg's Disease *(Rare)*

Diagnosis

Pain and tenderness are found under the second metatarsal head in an adolescent.

The X-ray may not show changes for several weeks.

Pathophysiology

Unknown. Falls into the category of the osteochondroses.

Treatment

Symptomatic measures (decreased activity, analgesics, a donut-shaped pad under the sore area) and usually all that is necessary. The problem resolves spontaneously in a few weeks to a few months.

Kohler's Disease *(Rare)*

Diagnosis

Pain, swelling and tenderness of the tarsal navicular bone are found in a prepubertal or pubertal child.

The X-ray may not show anything early in the course.

Pathophysiology

Unknown. An osteochondrosis.

INTOEING GAIT

Treatment

Advise the patient regarding *symptomatic measures* (decreased activity, analgesics, open shoes, or padding over the painful area).

The condition resolves of its own accord over several weeks to several months. Severe cases should be referred for consideration of casting.

HEEL CORD PAIN IN CHILDREN

Unless it is simply a pressure phenomenon from ill-fitting shoes, this is in- due to SEVER'S DISEASE.

Sever's Disease *(Uncommon)*

Diagnosis

Pain and tenderness occur at the achilles tendon insertion in an adolescent or preadolescent.

The X-ray may or may not show fragmentation at the calcaneal apophysis.

Pathophysiology

This is a traction-type injury at the apophysis where the achilles tendon attaches; analogous to Osgood-Schlatter's Disease in the knee.

Treatment

A ¼" *heel lift* in both shoes will be helpful. Calf stretching should be advised (Patient Handout 20-2).

Decrease activity to achieve a level of pain acceptable to the patient. Rest is mandatory if pain or swelling are severe.

Aspirin and ice may be used as needed.

The natural course of the condition is for it to disappear completely when the apophysis closes.

INTOEING GAIT

Evaluation

Observe the child and estimate the angle of gait, i.e., the angle the feet make with the line of progression. Normal is neutral to 30° of outtoeing.

Examine the foot for a METATARSUS ADDUCTUS deformity (Figure 19-1b). This condition was discussed earlier in this chapter.

266   19. PEDIATRIC PROBLEMS

The degree of TIBIAL TORSION can be evaluated by measuring the thigh-foot angle (Figure 19-3a). At birth, this is usually about zero degrees, though some infants are born with residual internal torsion. The tibia rotates outward approximately 20° by the time the child is walking. If abnormal, see below.

Measure hip rotation as shown in Figure 19-3b. The sum of external and internal rotation is usually about 100°. If internal rotation is greater than 70° excessive FEMORAL ANTEVERSION is present; if more then 80°, the abnormality is considered severe. (Femoral anteversion is the angle made between the femoral neck and a line between the distal femoral condyles. The normal child has 40° or so of femoral anteversion at birth, decreasing to 10° by about age eight.) See the discussion below.

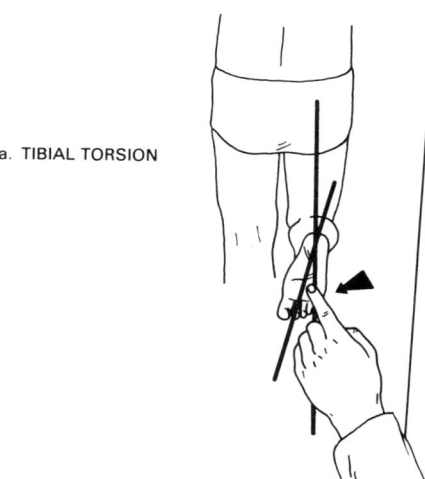

a. TIBIAL TORSION

Estimate the angle between the thigh and foot.

b. FEMORAL VERSION

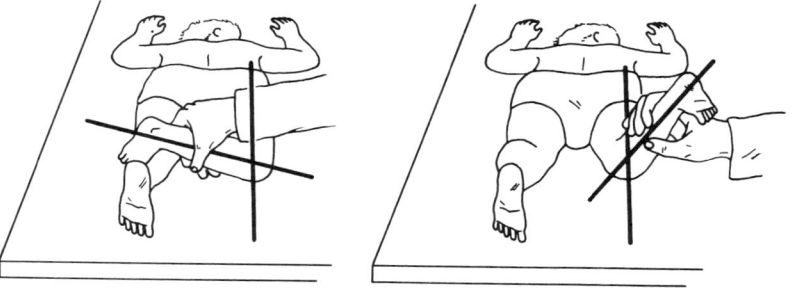

Estimate maximal internal rotation and external rotation (see text).

Fig. 19-3.   Evaluating tibial torsion and femoral version.

Internal Tibial Torsion *(Common)*

Diagnosis

As explained above.

Treatment

Although some authorities still recommend the use of a night brace, the trend is toward advising no treatment at all, except perhaps the avoidance of habitual sleeping and sitting positions which reinforce and possibly perpetuate the deformity.

Excessive Femoral Anteversion *(Uncommon)*

Diagnosis

As above.

Treatment

Since the natural course is for the abnormality to decrease with time, and for compensatory external tibial torsion to occur, no treatment is necessary unless by age eight there is still a gait abnormality severe enough to warrant the consideration of surgery.

BOWLEG, KNOCKKNEE, AND BACKKNEE

Evaluation and Management

Bowing of the legs (GENU VARUM; see Figure 19-4a) is normal until about the age of eighteen months, at which time it should begin to correct, and eventually may result in a knockknee deformity (GENU VALGUM; see Figure 19-4b), which in turn may normally persist to age five to seven. No treatment is required, except as noted in the following paragraphs.

If bowing does not begin to decrease by age two, or if it is severe or unilateral, an A-P X-ray of the legs should be performed to rule out rickets, Blount's Disease (osteochondritis of the medial proximal tibial epiphysis) or other metabolic or developmental conditions.

Similarly, if knockknee is severe (more than 15° or so) or unilateral, or does not begin resolving by age seven, radiographs should be obtained to rule out underlying disease.

Congenital GENU RECURVATUM (Figure 19-4c) is usually due to intrauterine malpositioning and is especially common in breech babies. Tibial sublux-

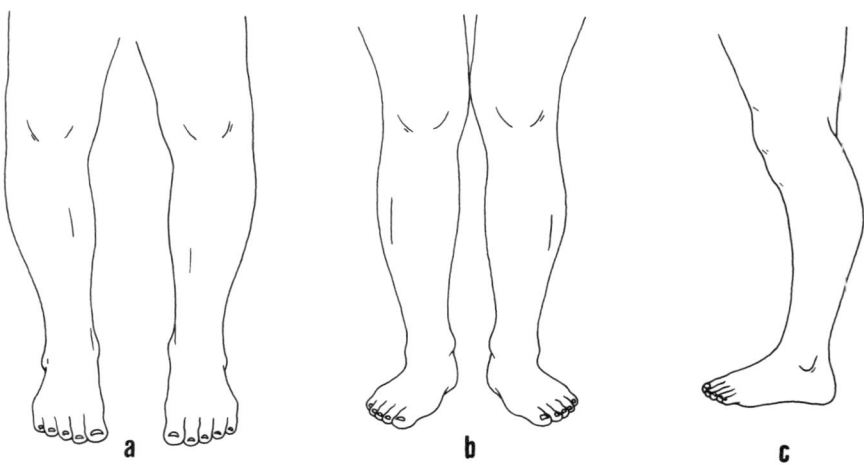

Fig. 19-4. Genu varum, valgum and recurvatum.

ation must be ruled out by X-ray; if it is, no treatment is necessary and the deformity usually self-corrects in a few weeks. If it persists, however, the infant should be referred to the orthopedist for splinting, etc.

KNEE PAIN IN CHILDREN AND ADOLESCENTS

Evaluation

Never forget that knee pain can be the sole complaint of a child with serious disease of the *hip*. Every complaint of knee discomfort mandates an examination of the hip.

An X-ray should always be done to rule out infection, neoplasm, or occult trauma.

Pain and tenderness, often with a lump and often bilateral, over the tibial tubercle in a preteen or adolescent suggests OSGOOD-SCHLATTER'S DISEASE.

Aching pain that is worse with activity (especially climbing hills or stairs), stiffness after sitting with the knee flexed, and painful grating on passive compression and movement of the patella against the femur suggest PATELLOFEMORAL PAIN. This condition is common in teenagers, especially girls, and is discussed in Chapter 15.

A history of episodes of knee collapse associated with lateral displacement of the kneecap suggests RECURRENT PATELLAR SUBLUXATION, a condition most often found in teenage girls. Refer to Chapter 15.

A history of collapse and/or vague pain may be found with patellofemoral pain, but also could be indicative of OSTEOCHONDRITIS DISSECANS or a MENISCAL TEAR. These conditions are also discussed in Chapter 15.

For further details about evaluation of the painful, collapsing, swollen, or injured knee, see Chapter 15.

Osgood-Schlatter's Disease *(Common)*

Diagnosis

Tenderness, often with a lump, will be found over the tibial tubercle (Figure 15-4) in a rapidly growing adolescent.

It is most common in athletically inclined boys.

X-ray may or may not show fragmentation at the tubercle.

Pathophysiology

This is a traction-type injury on the apophysis where the patellar tendon attaches; some hold that it is due to tendon growth not keeping up with bone growth.

The same process occurring at the other end of the patellar tendon (i.e., the inferior pole of the patella) is called Larsen-Johannson's disease.

Treatment

Decrease activity to achieve a level of pain acceptable to the patient. Rest is mandatory with severe pain or significant tenderness.

Aspirin and ice may be used as needed.

The natural course of the condition is for it to disappear completely when the apophysis closes.

HIP PAIN IN CHILDREN AND ADOLESCENTS

Evaluation

A child with hip pain or limitation of motion who has more than a low-grade fever, or who has an elevated WBC count or sedimentation rate, or who appears ill, must be considered to have a septic joint until aspiration proves otherwise. The septic hip is more often found in the infant (where a unilaterally flexed hip, refusal to move the leg, or crying with diapering may be the only clues to the cause of the child's systemic toxicity) but can occur at any age. Rapid destruction of the joint occurs if aggressive therapy is not instituted immediately; STAT consultation with the orthopedist is mandatory.

Ask about other joint involvement to help rule out juvenile rheumatoid arthritis.

Thomas' test (Figure 19-5) can reveal whether there is a flexion contracture of the hip, which is a sign of intrinsic hip disease.

An X-ray must be done to rule out osteomyelitis, tumor or occult trauma, as well as to look for signs of the three conditions discussed below.

Keep in mind that hip disease often manifests itself as a complaint of pain in the knee, abdomen or back instead of the more expected groin pain or limp.

Atraumatic hip pain in the child or adolescent, if the conditions mentioned above are ruled out, may be due to one of three diseases, all of which must be considered in the differential diagnosis.

Transient Synovitis of the Hip *(Uncommon)*

Diagnosis

Pain and limp may be of fairly sudden onset, or can come on gradually over one to two weeks.

There can be a low grade fever, but if there is significant elevation of temperature, white cell count or sedimentation rate, or if the child appears toxic or in much pain, a septic arthritis must be ruled out by referring the patient immediately for joint aspiration.

The condition is most commonly seen from ages three to six, but can occur up to age twelve.

On examination, the child will have a toe-walking gait and will hold the hip in flexion.

Range of motion will be limited and painful at the extremes. (In septic arthritis *any* motion will be painful.)

X-ray may be normal or only show evidence of a joint effusion.

Pathophysiology

The synovitis may be a result of unrecognized trauma or possibly of a viral infection.

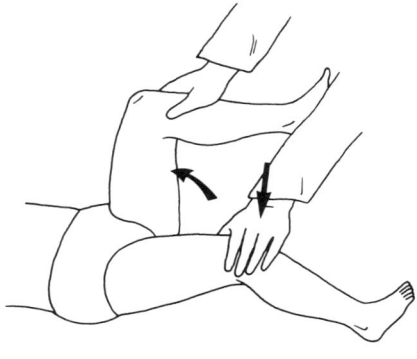

Fig. 19-5. Thomas' test. Flex the opposite hip and knee 90° to flatten the pelvis, then see if you can passively fully extend the hip you are testing. If not, a flexion contracture is present.

HIP PAIN IN CHILDREN AND ADOLESCENTS

Treatment

The child should be placed at *bedrest*, with *aspirin* used for pain.

Resolution usually over several days. When reexamination reveals pain-free range of motion and loss of the flexion contracture, weight-bearing may be resumed.

All children with this condition must be reexamined and X-rayed again after several months to rule out LEGG-PERTHE'S DISEASE (see below) which is eventually found in one-tenth of cases originally diagnosed as transient synovitis.

Legg-Perthe's Disease (Avascular Necrosis of the Femoral Head) *(Uncommon)*

Diagnosis

This can be a cause of hip pain in children three to twelve years of age, in boys more often than in girls.

Groin pain (sometimes referred to the knee or abdomen) and a limp may begin fairly abruptly but more commonly are noted to come on over weeks to months. Symptoms are exacerbated by activity and somewhat relieved by rest.

Examination reveals a flexion contracture and limitation of motion, with pain at the extremes.

X-ray is characteristic (showing increased density of the femoral head) but may remain normal for the first couple of months, and so must be repeated at intervals before the condition can be ruled out.

Sickle cell disease should be tested for in black children with this condition.

Pathophysiology

This is an avascular necrosis of the femoral head, the etiology of which is unknown and probably multifactorial. It is often classified with the osteochondroses, but whether it belongs there is a matter of conjecture.

The disease passes through three stages: necrosis, revascularization, and reossification.

Treatment

Proper management is complex and is critical to prevent severe hip problems later in life. The patient should be managed by an orthopedist.

Slipped Capital Femoral Epiphysis *(Uncommon)*

Diagnosis

It is most commonly found in ages ten to seventeen, especially in tall and this boys or conversely, in quite obese ones. One-fourth are bilateral.

The onset of pain in the groin (or often referred pain to the knee) and a limp may be sudden after trauma, or gradual over weeks to months.

On examination there will be a flexion contracture and also quite marked restriction of internal rotation.

A-P and lateral X-rays will usually show the slip, but if they do not and suspicion remains, a true lateral view should be ordered and may be revealing. (The femoral head normally looks like a scoop of ice cream on a cone; in epiphyseal slippage, the ice cream appears to be sliding off the cone.)

Pathophysiology

Hormonal factors are thought to play a role in weakening the epipyseal plate and allowing the slip posteriorly and inferiorly of the femoral head.

Treatment

This condition should be considered an *emergency* and referred to the orthopedist immediately for pinning.

BACK PAIN IN CHILDREN AND ADOLESCENTS

Evaluation

History and examination are as discussed in Chapter 11. Children can have ACUTE BACK STRAINS though much less commonly than adults. Discogenic low back pain is quite rare in childhood. These conditions are covered in Chapter 11.

The possibility of pain *referred* from the kidneys, other retroperitoneal structures, the abdomen, or a hip must be kept in mind, and conditions there must be ruled out by appropriate examination and laboratory tests.

An X-ray must be performed in all children with backache, especially to rule out osteomyelitis or tumor. If SPONDYLOLISTHESIS is found, see Chapter 11. Shmorl's nodes are diagnostic of SCHEUERMANN'S DISEASE, which is discussed below. Abnormalities in the curvature of the spine are discussed in the next section.

Unilateral lumbar pain worse with extension in a gymnast, dancer or other athlete who frequently arches his or her back suggests STRESS FRACTURE OF THE PARS INTERARTICULARIS, discussed in Chapter 11.

A child with persistent backache and a normal X-ray may well have DISCITIS or early OSTEOMYELITIS. A bone scan will be positive before the X-ray shows any changes; an elevated sedimentation rate is almost invariably found with these conditions. See Chapter 11.

Scheuermann's Disease *(Rare)*

Diagnosis

A complaint of back pain is given by an adolescent, and sometimes an increased kyphosis is seen on exam.

Shmorl's nodes are seen on X-ray.

Pathophysiology

This is an idiopathic breakdown of the developing vetebral end-plate.

The breakdown allows herniation of disc material into the vertebral body and can lead to an increasing kyphosis.

Treatment

Extension exercises and *postural instruction* are the mainstays. These patients must be observed carefully for the development of kyphosis which can lead to severe back problems later in life. At the first sign of increasing curvature the patient should be referred to a specialist for consideration of bracing, etc.

Otherwise, *rest* and *antiinflammatory medication* should be used as needed.

The disease becomes asymptomatic spontaneously.

ABNORMAL CURVATURE OF THE BACK

Evaluation

A child or adolescent with excessive kyphosis (roundback) should be referred to a specialist for management.

Any lateral curvature (SCOLIOSIS) is abnormal. A *functional* scoliosis is present if the curve disappears on recumbency. It may be due to leg length inequality, muscle contracture about the hip, or splinting secondary to back pain from any other condition. Correction is by treatment of the underlying cause. A *structural* scoliosis remains present even with a change in position, and is a much more serious matter. A screening examination for scoliosis as described below *must be performed yearly* on preteens and adolescents. (More and more schools are instituting such programs.)

Structural Scoliosis *(uncommon)*

Diagnosis

Lateral curvature does not disappear with recumbency.

An asymmetric rib hump or asymmetry of the shoulders, scapulae or lumbar muscle masses can be seen from behind when the patient bends forward with the hands together (Figure 19-6).

Flank creases may be asymmetric.

A plumb line dropped from the center of the cervical spine misses the gluteal cleft. (This test may fail in a compensated double-curve scoliosis.)

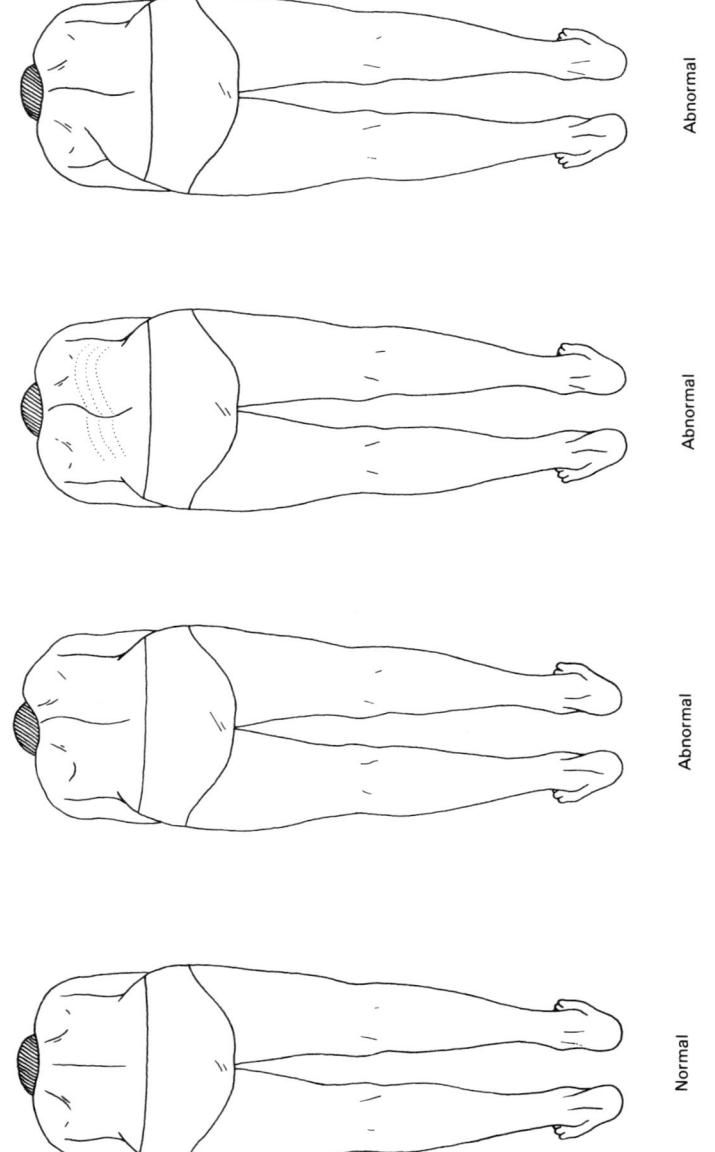

Fig. 19-6. Forward bend test for scoliosis.

Pathophysiology

Most cases are idiopathic and thought to be genetically determined, but specific vertebral malformations or neuromuscular disease may sometimes be responsible.

Treatment

Proper management is essential to avoid possible skeletal, neurologic and cardiorespiratory complications. All children and adolescents with any degree of structural scoliosis noted on examination or X-ray must be *referred* to an expert in the field for follow-up.

Once epiphyses have closed, progression is not likely unless the degree of curvature is great; over the age of 25 only patients with more than 25° of scoliosis need be referred. The management of back pain in those with lesser curves is similar to that of other patients with back pain.

WRYNECK IN INFANTS AND CHILDREN

Evaluation and Management

Infants and young children with a *fixed* lean of the head to one side and rotation toward the other side have a CONGENITAL MUSCULAR TORTICOLLIS (unilateral contracture of the sternocleidomastoid muscle). If the deformity is mild and there is no mass palpable in the sternocleidomastoid muscle, advise the parents to do the following:

Position the infant for sleep prone, with the head turned to alternate sides.

Position him seated so that he must turn his head to stretch the affected muscle in order to see what is going on.

Gently stretch the tight muscle at each diaper change.

If the deformity is severe or a mass is palpable, the child should be referred to the specialist for treatment. Note that a fair percentage of these children also have congenital hip dysplasia.

Sudden atraumatic onset of pain on one side of the neck with deviation of the neck (ACUTE TORTICOLLIS) is managed as in adults. See Chapter 4.

ELBOW PROBLEMS IN CHILDREN

The pulled elbow in a toddler and elbow pain in the child who does much throwing are the only common afflictions of this joint (aside from fractures) in the pediatric age group.

Radial Head Subluxation (Pulled Elbow, Nursemaid's Elbow) *(Uncommon)*

Diagnosis

The injury usually results from a toddler one to four years of age being pulled or swung by his or her extended arm, as when a parent tries to help the child negotiate a curb.
The child holds his forearm pronated and his elbow flexed.
The X-ray is normal.

Pathophysiology

The head of the radius subluxes through the annular ligament, which normally holds it in place.

Treatment

The subluxation is *reduced* by grasping the child's forearm with one hand while pressing the thumb of your other hand over the radial head; the forearm is supinated and the elbow flexed, and the subluxation reduces with a palpable snap. After a few moments the child is pain free and no furthur treatment is necessary.

With recurrent subluxations, the use of a *sling* for a few weeks is helpful.

Overuse Injuries Related to Throwing
(Little League Elbow) *(uncommon)*

Diagnosis & Pathophysiology

In a patient that overuses the arm as in pitching, several syndromes can commonly occur:

- muscular strains about the shoulder
- rotator cuff impingement, inflammation and breakdown (more common in teenagers and adults)
- stress fractures of the humerus (don't miss these! Bone scan may be necessary.)
- traction apophysitis of the medial epicondyle (in the immature skeleton, can progress to a true fracture, with subsequent growth disturbance and deformity; in adults, the traction stress at the medial elbow more often leads to tendonitis or collateral ligament breakdown)
- ulnar nerve inflammation and dysfunction
- osteochrondritis of the radiohumeral joint (secondary to repetitive compression forces. Any pain or tenderness in the anterior or lateral elbow is serious business and should be referred for specific diagnosis and management. Joint deformity may result.)
- inflammation and breakdown in the posterior elbow (triceps insertion).

Management:

Mild syndromes should be treated with:

- *rest* until symptoms abate
- gradual resumption of activity to a level that doesn't lead to pain
- proper *warmup* before throwing
- *ice* application afterward
- and, probably most important to prevent recurrence, teaching of proper *mechanics* by a *qualified* pitching coach.

Moderate to severe symptoms, or any arm with a flexion contracture,* positive radiologic findings, radiohumeral joint pain, neurologic symptoms, or midhumerus pain should be managed with total rest and orthopedic referral.

MUSCULOSKELETAL TRAUMA IN CHILDREN

We here reiterate three of the principles discussed in the opening of this chapter and add another.

Any periarticular trauma in a child must be considered to be a possible epiphyseal plate injury. These can have serious consequences, and point tenderness over an epiphysis may be the only finding. (X-rays can be normal.)

Comparison views of the uninjured side should always be taken.

Multiple or repeated trauma or a history not quite consistent with the physical findings could mean the *Battered Child Syndrome*. This is a potentially life-threatening situation. Failure to make the diagnosis and act appropriately on it, whether due to oversight or to fear of being wrong and offending the parents, is a mistake that must be avoided.

Always check neurovascular status distal to any injury.

Tuft fractures of the distal phalanges and fractures of the small toes are managed as in adults (see Chapter 9 and 16). TORUS FRACTURES of the forearm are discussed below. Dislocation of the radial head was discussed above. All other fractures and dislocations in children are beyond the scope of this book; the reader is advised to refer the patient to an orthopedist or to consult an orthopedic text or manual. Bruises and myositis ossificans are discussed in Chapter 21.

---

*i.e., lack of full extension; a non-specific sequel of significant elbow inflammation.

Torus Fractures of the Distal Radius *(Common)*

Diagnosis

X-ray reveals a buckling of one or both sides of the cortex of the distal radius, usually after a fall on the outstretched hand.

If there is tenderness over an epiphysis, or if on any view there is a discontinuity of cortex, the patient has more than a torus fracture and treatment is more complex.

Treatment

Protective splinting for three to four weeks and a follow-up X-ray to assure healing are the only treatments necessary for a torus fracture.

SYSTEMIC ILLNESSES

Juvenile Rheumatoid Arthritis *(Rare)*

Clinical Features

There are three fairly distinct types:

1. *Still's Disease* is an illness in which systemic manifestations often overshadow joint complaints. There are daily fever spikes with in-between returns to normal temperature, a faint salmon-colored rash, lymphadenopathy, and sometimes hepatosplenomegaly. The white cell count and sedimentation rate are increased.

2. *Pauciarticular arthritis* involves a few large joints and is usually seen in younger children. This is the most common type. About one-fourth of these patients will develop potentially serious eye disease.

3. *Polyarticular arthritis* is seen with greater frequency as children grow older, and is almost certainly the same disease as adult rheumatoid arthritis.

Pathophysiology

The etiology of each of the three types is unknown. Joint pathology is analogous to that of adult rheumatoid arthritis.

Treatment

Treatment modalities are analogous to those used in adult rheumatoid arthritis. Growth retardation can result near severely affected joints. Referral to a rheumatologist or pediatrician is recommended. Children with pauciarticular disease must also be referred for ophthalmologic examination at least twice yearly, even if asymptomatic.

SYSTEMIC ILLNESSES

Rheumatic Fever *(Rare)*

Clinical Features

The arthritis of rheumatic fever characteristically is migratory and involves at least several joints; the originally involved joints become asymptomatic as other joints become inflamed. Large and medium-sized joints are most commonly affected.

Jones' criteria for the diagnosis of rheumatic fever are listed below. Evidence of preceding streptococcal infection (a positive throat culture or increased anti-streptococcal antibodies) plus two major (or one major and two minor) criteria indicates the probable presence of rheumatic fever.

Pathophysiology

The mechanism by which streptococcal infection sometimes leads to rheumatic fever remains unknown.

Treatment

Antiinflammatory medications, antibiotics, rest and *specific treatment for complications* (such as congestive heart failure) are used. The reader is referred to textbooks of pediatrics or internal medicine for details.

Table 19-2. Jones' Criteria for the Diagnosis of Rheumatic Fever

Major	Minor
Carditis	History of previous rheumatic fever
Polyarthritis	Evidence of rheumatic heart disease
Chorea	Arthralgias
Erythema marginatum	Fever
Subcutaneous nodules	Increased WBC count and ESR
	ECG changes

PATIENT HANDOUT 19-1

Stretching Your Infant's Foot

Your child has a mild foot deformity called *metatarsus adductus*. This usually gets better with time; you can help speed up the process by stretching the foot as shown below ten times at each diaper change.

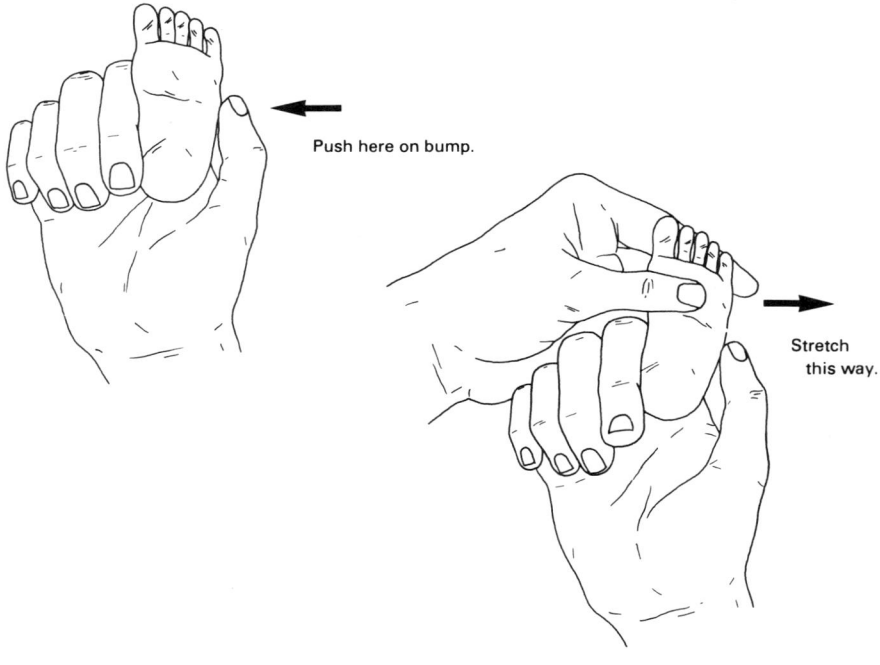

PATIENT HANDOUT 19-2

Your Baby's or Child's Shoes

Unless your child has a foot deformity for which the doctor has made specific recommendations, he or she needs shoes for only one reason: protection from injury. Your child should go barefoot whenever it is safe: when it isn't, dress him or her in shoes that are large enough, soft, flexible and well ventilated. The sole should be flat: high tops are helpful only in making the shoe hard to remove. Remember, a rigid shoe will slow the development of your child's foot muscles.

20

Running Injuries

Introduction 282
Overuse Injuries 283
Training 284
Grades of Overuse Injuries 284
Stress Fractures 284
Preventing Overuse Injuries 285
Principles of Treatment 286
    Allowing Healing 286
    Decreasing Inflammation 286
    Correcting Stress Concentrators (Ground Reaction Forces, Muscle Tightness,
        Hyperpronation, Other Factors) 287
Evaluation 288
    History 288
    Examination 289
The Biomechanics of Running 289
Skin Problems 289
Toe Problems 291
Differential Diagnosis of Runners Injuries 291
Treatment of Runners' Injuries 296

More and more Americans are running longer and longer distances, and many are paying a musculoskeletal price for their psychological and cardiovascular benefits. Those who see doctors not attuned to the biomechanics or the psychology of running are likely to be told simply to stop or cut down on their avocation, or possibly offered medications or injections; they are not likely to be offered advice about their running style, training method, or shoe that can relieve the problem or at least prevent its recurrence when natural healing has taken place and they are ready to resume their activity. For many runners this is unacceptable, and they will seek advice from podiatrists, other doctors, or sports medicine specialists, or sometimes, unfortunately, from unqualified friends or from quacks. This chapter is an effort to change that, so that primary physicians can give sound and acceptable advice to their running patients based on an understanding of the pathophysiology of common running injuries.

While the diagnosis and treatment of these problems are not overly complicated, they are not blatantly simple either. Even some writers of articles and books on running medicine commonly lump disparate problems under titles like "shinsplints" or "runner's knee," for example, and then prescribe the same treatment for all runners with leg pain or with knee pain. This must be recognized as an oversimplification that will often lead to suboptimal therapeutic advice.

A very, very important note: the following material will, hopefully, lead the reader into a mode of thinking about runner's injuries that will be logical, consistent and very practical. But those providing health care to runners must never forget that their patients are subject to the same diseases and conditions as others. A good biomechanical approach to their lower extremity complaints must not lead to overlooking such conditions as vein thrombosis, meniscal tears in the knee, and osteomyelitis, for example. Some such conditions can obviously be of much greater importance than the musculoskeletal overuse conditions described below.

OVERUSE INJURIES

The conditions described in this chapter are overuse injuries. It is extremely important in keeping one's thought processes clear and logical to differentiate clearly between this type of problem and an acute traumatic injury. The latter includes problems such as ankle and knee sprains, contusions and various muscle strains which occur suddenly when an acute force is applied; the patient steps in a hole, or is hit by a train, or feels something suddenly snap. The moment of injury is well defined. Treatment of these "macrotraumatic" or "acute overload" injuries includes protection of the injured tissue until healing can occur (sometimes after surgical or manipulative intervention); controlling swelling, inflammation and pain; and maintaining and then restoring flexibility and strength ("rehabilitation").

Overuse injuries, on the other hand *occur over a period of time when the forces applied to a structure are increased faster than the structure can adapt, or exceed its limits of adaptation.* (All living tissue will adapt itself to increased requirements; but all tissue also has a maximum rate at which it can adapt, and a final limit to its adaptation. These limits depend on a multitude of factors, usually hard to quantify. Previous injury will lower these limits to some degree, a situation that is commonly seen.)

TRAINING

In a very real sense, *training is a controlled overuse injury*. In order for musculoskeletal structures to become stronger, they must be stressed, but in such a manner that they can recover and strengthen before being restressed. When the stress on a particular structure is such that this cannot occur, it begins to break down - an overuse injury. Which particular structure begins to break down first is a function of the many factors that can serve to concentrate stress in a particular area: things like inadequate muscle flexibility, strength imbalance of opposing muscle groups, bony anatomy, terrain, running shoe, scarring or other damage from previous injury, etc., etc.

GRADES OF OVERUSE INJURY

The severity of an overuse injury can be graded according to the constancy of the pain:
GRADE I
Pain only after activity.
GRADE II
Pain starts during the activity.
GRADE III
Pain still persists the next day.
GRADE IV
Constant pain.

STRESS FRACTURES

A stress fracture is nothing more or less than an overuse injury to bone. Bone, like other living tissue, strengthens in response to applied stresses, but if the stress is too great or applied too rapidly it may begin to break down. In runners the stress may be from transmitted shock (''ground reaction force'') or from muscle (at an origin or insertion). Often *muscle fatigue* is what allows ground reaction forces to rise to harmful levels or leads to a strength imbalance which applies extra stresses to bone.

The diagnosis can be made by localized or significant bony tenderness. This will indicate at least an early stage of excess stress (''stress reaction''); if the patient does not cease the offending activity or make other adjustments, the process will advance to a ''stress injury'', where there is actual bony resorption, and then to a ''stress fracture'', where ''microfractures'' have occurred through the weakened bone. Obviously there is no clear dividing line between these stages.

Continued forces applied to a stress fracture may lead to an actual through and through "completed" fracture.

X-rays will often be negative early in the course of a stress fracture, and may never turn positive at all in milder degrees of bony overuse injury. If they do eventually turn positive, the findings can sometimes be quite subtle; sometimes only by correlating the point of maximum tenderness to a vague radiographic finding will one be convinced of its significance. The X-ray picture may be only of callus (thickening along the periostium or endostium), or a small sclerotic line, or a small clear fracture line (especially when transverse, this is the most likely kind to complete).

A bone scan can make the diagnosis of stress fracture very early in the course. It can be used if the diagnosis is in doubt, but is most useful in ruling out stress fractures in people who have pain in areas where the bone is not accessible to palpation: the proximal tibia, and especially the femur and pelvis in patients with groin, buttock or thigh pain.

Common overuse injuries in runners are those of the metatarsals, calcaneous, fibula, tibia, femur and pelvis (the latter mainly in female athletes). Stress fractures of the pars interarticularis of a lumbar vertebra are common in teenage gymnasts, dancers and football linemen; and various sites in the upper extremities can be involved in pitchers, tennis players and weight lifters.

PREVENTING OVERUSE INJURIES

The main way to prevent overuse injuries is quite simply to increase stress on the musculoskeletal system slowly enough to allow it to adapt. No one can dictate a training schedule in advance for themselves or others; activity should be increased gradually, such that no particular area of the body is consistently more sore than others, and so that there are no areas ot persistent pain. In other words, the athlete in training must listen to his or her body, and must be aware that there will be limits beyond which breakdown will occur. Pushing those limits without going past them is what training is all about.

It is best to run every other day rather than daily, or at least to alternate heavy workouts with light days. This allows more time for the tissues to recover from microtrauma before being stressed again. Such a schedule is especially important when recovering from injury.

Runners should also be sure to warm up and stretch properly before setting off, and select shoes which cushion shock adequately (i.e., running shoes, not tennis or other types of shoes). They should be aware of how hard, hilly or banked terrain can increase musculoskeletal stresses. If they excessively pronate or have other biomechanical "variances from the norm" they must also know how to minimize these effects. All these factors are covered in more detail in the discussion of treatment which follows.

PRINCIPLES OF TREATMENT OF OVERUSE INJURIES

For any given problem, various textbooks may list a bewildering array of different treatments. But actually all of them fall into three basic categories, and keeping these three fundamental classifications mentally separate will clarify the situation and make the process of giving advice to a runner with an injury much more pleasant and effective.

The three categories are simply:

1. Adjust activity schedule to allow healing.
2. Reduce inflammation.
3. Correct stress concentrators, i.e., those biomechanical or external factors which have placed inordinate stress on the particular part that eventually broke down, i.e., the site of injury.

Allow Healing

Training must be cut back enough so that, while the other measures listed below are applied, things are getting better, not worse.

Three exceptions to this general rule:

1. In cases of overuse injury to bone, i.e., "stress injury" or "stress fracture", activity should be completely discontinued until pain and tenderness are gone. In some femoral and in severe tibial stress fractures, even further protection may be necessary, i.e., avoiding weight bearing altogether, casting, etc.
2. In cases of *severe* inflammation of soft tissue structures total abstention is also a good idea. The achilles and patellar tendons, among others, are liable to suddenly tear if stress is continued when they are very inflamed. And if there is enough pain to affect gait, there is real risk of an acute traumatic injury or an overuse injury to another structure.
3. If symptoms are prolonged there is a real risk of developing chronic irreversible changes (scarring) which will forever "lower the limits" of the injured structure.

If running must be completely discontinued for a time, substitute aerobic activities such as swimming or (if safe) cycling may be performed.

Decrease Inflammation

A variety of measures are useful for their effects in reducing inflammation.

1. The most basic, and one that is useful in nearly all cases, is *ice*. It should be applied to the injured area for about twenty minutes at a time; long enough to

PRINCIPLES OF TREATMENT OF OVERUSE INJURIES

get by the initial moments of increased pain and to obtain significant tissue cooling, but not long enough to cause any actual tissue damage.* Ice should be applied after activity and as needed.
2. Oral medications may include:
   - aspirin, taken one-half hour before activity and prn, or on a more regular schedule
   - prescription NSAI's
3. Cortisone injections may be resorted to where not contraindicated.
4. Physical therapy modalities such as ultrasound, electrical stimulation or contrast baths may be helpful.

Correct Stress Concentrators

There are a variety of different factors which cause force overload in one particular area or another; which to look for and try to correct depends on the particular structure injured, and is discussed along with the specific syndromes later in the chapter. Common factors are:

Ground Reaction Forces, or *Transmitted Shock.*

This is the prime stressor in many overuse injuries, and is at least a contributing and magnifying factor in most. It is more of a problem in runners with rigid feet. To minimize transmitted shock, advise the following:

- choosing shoes which absorb shock well, and obtaining new ones when they wear out (only a few hundred miles for most shoes; often the cushioning ability is lost long before the outsole wears out or other breakdown is evident)
- using extra cushions such as Sorbothane® viscoelastic full length inserts
- avoiding downhills as much as possible
- avoiding hard surfaces as much as possible
- correcting running style (a good coach can be helpful here) and lowering speed
- working on muscle flexibility (i.e., adequate warm-up and stretching) probably helps too.

Muscle Tightness

This should be corrected with adequate warm-up and then static stretching before activity, and again afterwards. See Patient Handout 20-2.

Hyperpronation

As discussed in the box on "biomechanics of gait", there is a continuous variation, depending on ligamentous tightness, subtalar angle, and other anatomic

---

*It is contraindicated over areas of decreased tissue perfusion, decreased sensation and open wounds.

factors, in the amount of foot pronation that occurs during gait. Runners who pronate more than average absorb shock well, but put more than average stress on various structures and are thus predisposed to various injuries (see Table 20-1). Excessive pronation should be corrected if and only if it is causing a problem, i.e., one of the overuse conditions listed; otherwise adjusting biomechanics that the patient is functioning well with can itself lead to problems. To minimize pronation advise the following:

- avoiding inclined terrain, like the side of a banked road. This is often a factor when hyperpronation-related symptoms are unilateral and examination of the patient doesn't indicate a lot of excessive pronation. An equivalent situation occurs when one leg is longer and that foot is forced to pronate more.
- choosing shoes that are designed to help minimize pronation. Features to look for include: a snug, well fitting and firm heel counter (a *reinforced* heel counter is useful to prevent premature breakdown); construction on a curved instead of a straight last; good arch support; and areas of different compressability in the midsole to minimize motion.
- adding a medial heel wedge to the shoe. See Figure 18-3. (There is a slightly increased risk of ankle sprains but a small wedge added to a flexible foot is usually safe). An over-the-counter arch support may also be helpful.
- In cases which don't adequately respond to these simple measures, it may be necessary for patients to obtain customized "orthotics" to correct their biomechanics. Referral to a podiatrist is best; be aware that these can be expensive. Note that orthotics are *not* indicated in cases where the overuse injury is not due to hyperpronation or to another biomechanical problem in the foot; nor as a preventative measure (they may cause trouble by upsetting well-compensated biomechanics); nor to increase efficiency of running (they don't).

Other Factors

Depending on the specific problem, other corrections may be advisable, such as more or less flexible soles, raised or lowered heels, correcting strength and balance between antagonistic muscles. partially correcting leg length discrepancies, and relieving shoe pressure points.

EVALUATION

History

Important points to question the patient about include:

- any specific trauma or strain
- change in schedule, terrain or shoe that may have precipitated the symptoms
- previous injuries especially to the involved area
- points specific to the involved joint or area
- enough information to rule out non-overuse problems as the cause of the patient's symptoms.

SKIN PROBLEMS

Table 20-1. Hyperpronation-Related Overuse Syndromes in Runners

Morton's Neuroma (some)
Metatarsal Stress Fracture (some)
Plantar Fasciitis (many)
Achilles Tendonitis (many)
Medial Distal Tibial Syndrome (almost all)
Tibial Stress Fracture Type a (almost all)
Patellofemoral Pain (few)

Examination

- look for tenderness. Bony tenderness suggests stress fracture.
- if a joint is involved, examine for range of motion and stability.
- look for joint effusion or soft tissue swelling.
- watching the patient's gait can give a a good indication of the patient's functional status and of the severity of the injury.
- measure leg lengths (see Figure 11-3).
- examine the patient's foot and shoes to determine how much the patient pronates (see Figure 20-1).
- do whatever other examination is mecessary to rule out a non-overuse etiology for the patient's complaints.

The Biomechanics of Running

Please review the "biomechanics of gait" in Chapter 18. The major differences with regard to the foot in running, as opposed to walking, are: (1) in running there is a phase in which both feet are off the ground, leading to much greater transmitted shock at foot strike (200-300 percent body weight as opposed to around 85 percent body weight when walking), magnifying stresses considerably; (2) in running, at the time of heel strike the foot is somewhat more supinated and *then* moves into pronation (to a varying amount and at a varying speed depending on anatomy, shoe and terrain) before moving back into supination for push off; (3) a fair percentage of runners, especially higher level athletes, actually make first ground contact with their forefoot rather than their heel. These runners often need extra forefoot cushioning in their shoes.

SKIN PROBLEMS

For a discussion of calluses and corns, see Chapter 18.

Blisters *(Extremely Common)*

Diagnosis

By inspection. They are usually related to new shoes or marked increases in mileage.

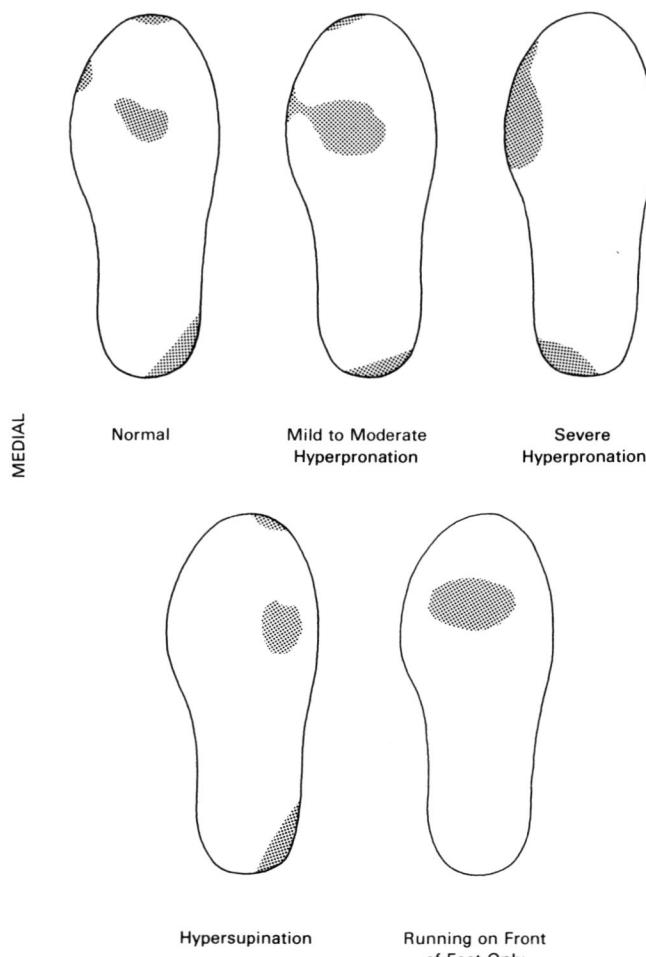

Fig. 20-1. Shoe wear patterns in runners.

Pathophysiology

Localized shearing stresses lead to separation of the skin layers, with fluid extravasation between them.

Treatment

Large blisters should be *punctured* (not denuded) and *drained* using sterile technique.

Open blisters must be conscientiously kept *clean* to prevent infection.

Small or incipient blisters should be covered with *adhesive tape* to protect them form shearing forces.

DIFFERENTIAL DIAGNOSIS OF RUNNERS INJURIES

Well-fitting shoes worn without socks will help prevent the development of blisters.

TOE PROBLEMS

Runner's Toe *(Common)*

Diagnosis

There will be inflammation and sometimes ecchymosis, especially near the nail bed.

Pathophysiology

Most commonly it is due to shoes that are too short with subsequent axial pressure on the nails.

Sometimes the problem can be a consequence of a habit or running with the toes curled, with the tips being pounded into the bottom of the shoe; this running style can be a consequence of wearing shoes that are too *big* or of coming downhill.

Treatment

Be sure *shoes* are of the proper length; one to two cm beyond the tip of the toes when standing is about right.

Nails obviously must be kept *well-trimmed*.

If the patient persists in running with his toes curled, temporary placement of a pad over the dorsal PIP joints of the involved toes can help cure him of the habit.

DIFFERENTIAL DIAGNOSIS OF RUNNERS INJURIES

The following is broken down by area of presentation. Often more information will be found in the corresponding chapter earlier in the book. Treatment for each syndrome is summarized in Table 20-4.

Pain in the Ball of the Foot

See figure 18-6. A hallux valgus deformity and bunion is treated as described in Chapter 18. Diffuse joint line tenderness, pain and sometimes crepitus and restriction of range of motion implies DEGENERATIVE ARTHRITIS (SYNOVITIS) OF THE FIRST MTP JOINT. Localized tenderness under a sesamoid with a positive compression test indicates SESAMOIDITIS.

Pain in the Transverse Arch

A lancinating pain, often worse with tight shoes and better barefoot, along with findings of tenderness *between* two metatarsal heads and sometimes reproduction of pain with lateral squeezing of the forefoot, implies a MORTON'S NEUROMA. Tenderness directly under a metatarsal head, often with callus, indicates METATARSALGIA. See Figure 20-2.

Pain in the Top of the Foot

Usually indicative of a METATARSAL STRESS FRACTURE, especially with localized bony tenderness. (Stress fractures are discussed earlier in the chapter). Occasionally due to EXTENSOR TENDONITIS, if tenderness is localized to a tendon and pain reproduced with passive stretch or resistance of that tendon.

Pain in the Bottom of the Heel and Arch

Most commonly PLANTAR FASCIITIS; occasionally a HEEL PAD CONTUSION or entrapment of a branch of the calcaneal nerve. See Figure 20-2.

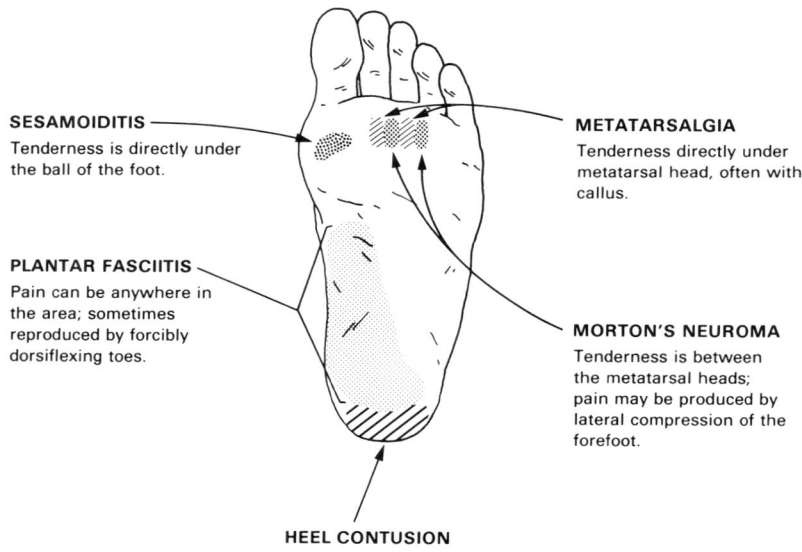

Fig. 20-2. Pain in the bottom of the foot.

DIFFERENTIAL DIAGNOSIS OF RUNNERS INJURIES

Pain with squeezing the heel suggests a calcaneal stress fracture (uncommon in runners).

Pain Around the Ankle

See Figure 17-2. POSTERIOR TIBIAL, ANTERIOR TIBIAL, or PERONEAL TENDONITIS are often responsible; occasionally the pain is from within the ankle joint (degenerative arthritis or previously undiagnosed osteochondral fracture) or the subtalar joint.

Pain in the Achilles Tendon Area

See Figure 17-9. ACHILLES TENDONITIS is very common, with tenderness at its insertion (usually the medial side) or several cm. proximally.

Pain in the Calf

Always rule out deep vein thrombosis, though a CALF STRAIN is more likely. See Chapter 16.

Pain in the Shin

See Figure 20-3.

Tenderness along the MEDIAL DISTAL TIBIA is what is usually called "shinsplints" and is a tendonitis/periostitis at the origin of the posterior tibial or medial soleus muscles, often related to hyperpronation, and quite common. The development of severe pain or significant or localized tenderness usually indicates that the stress on the bone has progressed to where a TIBIAL STRESS FRACTURE is present. (A bone scan shows diffuse linear uptake in the former, more spotty uptake and eventually "hot spots" in the latter).

Bony tenderness elsewhere or deep, poorly localized pain raises the probability of a STRESS INJURY or FRACTURE of the tibia or fibula.

Pain along the anterior compartment, especially if associated with transient weakness (in toe or foot dorsiflexion) or numbness (in the first web space), indicates a RECURRENT ANTERIOR COMPARTMENT SYNDROME, which may also present as, or progress to, an acute compartment syndrome requiring STAT orthopedic referral.

Always check distal neurovascular status.

Pain in the Knee

Knee pain in runners is usually caused by the syndromes that follow, but not uncommonly can be due instead to meniscal tears or degenerative arthritis, espe-

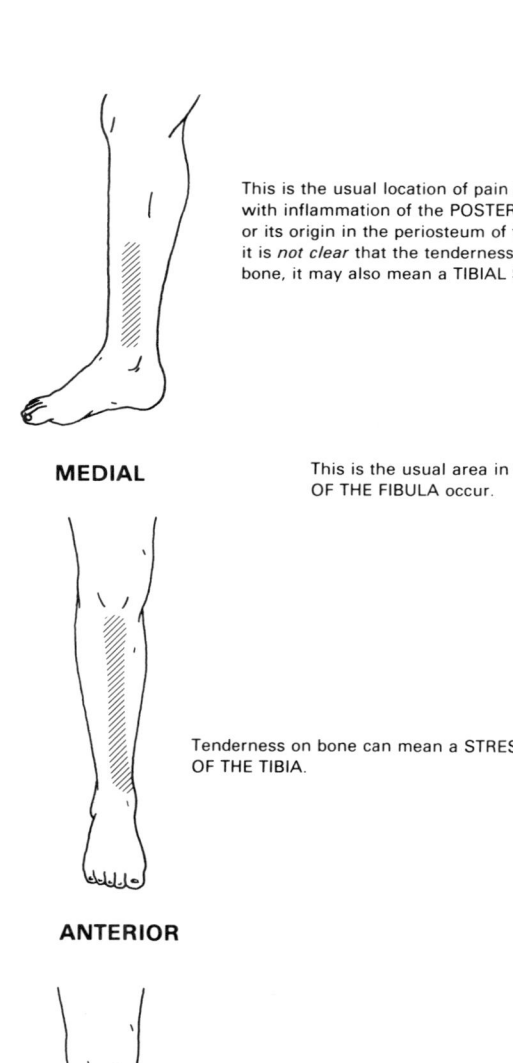

Fig. 20-3. Pain in the shin.

DIFFERENTIAL DIAGNOSIS OF RUNNERS INJURIES

cially in the previously injured or older runner. Joint line tenderness, knee effusions or mechanical symptoms such as catching or giving way are clues; refer to Chapter 15.

More usual, however, is an anterior knee ache, sometimes with crepitus and "almost" collapsing, usually signifying PATELLOFEMORAL PAIN ("runners knee"). Pain and tenderness over the patellar tendon or iliotibial band indicate PATELLAR TENDONITIS or ITB TENDONITIS, respectively. Posterior pain is usually due to a CALF STRAIN, HAMSTRING STRAIN, or rarely, Baker's Cyst, or may be a symptom of sciatica. Remember that knee pain can be referred from the back or hip, and that bony tenderness may indicate stress fracture.

Pain in the Thigh

Lateral thigh pain is usually due to ITB TENDONITIS or TROCHANTERIC BURSITIS. Pain and tenderness in the posterior or medial thigh usually indicates HAMSTRING STRAIN or ADDUCTOR STRAIN respectively. Quadriceps strains are rare in runners. Keep in mind the very real possibility of a FEMORAL STRESS FRACTURE, which may be hard to diagnose clinically; bone scan should be performed with the slightest suspicion (if the tender area cannot be palpated, e.g.).

Groin Pain

STRESS FRACTURE OF THE FEMUR (especially femoral neck) must first be excluded by bone scan before further activity is permitted or other diagnosis made; tenderness and pain with the appropriate passive and resisted motions suggests a GROIN STRAIN (iliopsoas, rectus femoris or adductor). Also rule out hernia, adenopothy or other mass, or pain referred from genital or abdominal structures.

Buttock Pain

Usually a hamstring strain, but may be referred pain from the back of hip; the latter especially must be ruled out.

Iliac Pain

Usually a strain of the Tensor Facia Lata (origin of the iliotibial band) or other muscle origin; can be an avulsion type stress fracture, especially in teenagers.

Unilateral Neck Pain

This is often due to running on an inclined surface, with the neck tilted to compensate.

TREATMENT OF RUNNERS' INJURIES

See Tables 20-2 through 20-4 immediately following.

Table 20-2.   Measures to Reduce Ground Reaction Force (Transmitted Shock)

Major Factors:
1. Avoid downhills
2. Avoid hard surfaces
3. Reduce speed
4. Choose shoes with adequate cushioning *(especially in rear foot for heel strikers, in sole for mid foot strikers)
5. Be sure shoes haven't lost their cushioning (midsole often breaks down after a few hundred miles, before other evidence of wear apparent)
6. Add visco elastic padding*

Other Factors:
7. Adjust running style (coaching)
8. Good muscle flexibility (warm up, stretching)

*Be aware that increased cushioning in the heel area, by allowing more calcaneal motion, *increases* pronation; a trade-off.

Table 20-3.   Measures to Control Hyperpronation*

1. Avoid inclined surfaces
2. Choose shoes with the following characteristics:
   - snug and stiff heel counter
   - (reinforced heel counter helps prevent premature breakdown)
   - built on curved last
   - good arch support
   - increased midsole density medially
3. Add medial heel wedge**, off the shelf arch support
4. Customized orthotics

*Be aware that more motion control means less shock absorbed within the foot and more transmitted up the leg; a trade-off.
**Can predispose to ankle sprain.

Table 20-4.   Treatment of Runner's Overuse Injuries

For all advise (except as noted):
1. Decreasing activity enough to allow symptoms to resolve, alternate day schedule.
2. Proper warm up and stretching.
3. Ice for twenty minutes after activity and as needed.
4. Aspirin one half hour before running, or more frequently if needed, or a course of prescription NSAI'S; possibly physical therapy modalities if needed.
5. Specific treatment as shown.

Table 20-4. Treatment of Runner's Injuries *(cont.)*

Ball of the Foot (DDx on pg. 29)

Synovitis (DA) of one MTP joint	• Rigid sole (be careful not to cause achilles problems) • Avoid hills, adequate padding under sore area	• Steroid injection may be offered (Figure 18-7) • Surgery last resort (Silastic joint)
Sesamoiditis	• Avoid hills • Accommodative padding	• Steroid injection may be offered

Transverse Arch (DDx on pg. 292)

Mortons Neuroma	• Be sure shoes have adequate width • Reduce any hyperpronation (Table 20-3) • Adequate forefoot cushioning	• Steroid injections may help: (Figure 18-9) • Oral meds, ice not too helpful • Surgery may eventually be needed; usually good results
Metatarsalgia	• Accommodative padding • Correct hyperpronation (Table 20-3)	• Difficult problem • Surgery a last resort

Top of Foot (DDx on pg. 292)

Metatarsal Stress Fracture	• Must *discontinue* until pain and tenderness gone	• When resume: avoid hills, reduce shock (Table 20-2), correct hyperpronation if present (Table 20-3)
Extensor Tendonitis	• Be sure shoe pressure not contributing	

Bottom of Heel and Arch (DDx on pg. 292)

Plantar Fasciitis	• Avoid uphills and speed work • Use visco elastic heel pad • Calf stretching program • Correct any hyperpronation (Table 20-3)	• Use arch support in cavus foot • Steroid injection may help (Figure 18-5) • Surgical release a last resort, good success rate
Heel Pad Contusion	• Visco elastic heel pad • Reduce shock (Table 20-2)	

Ankle (DDx on pg. 293)

Posterior Tibialis Tendonitis	• Correct hyperpronation (Table 20-3) • Avoid shoe pressure on area	
Anterior Tibial Tendonitis	• Watch for pressure from laces • Avoid down hills	• Lower heel shoe
Peroneal Tendonitis	• Avoid shoe pressure on sore area • Stay off inclined surfaces	• Lateral heel wedge (Figure 18-3)
Subtalar Joint Synovitis	• Reduce shock in rigid foot (Table 20-4)	• Consider steroid injection

(continued)

	• Correct hyperpronation in flexible foot (Table 20-3)	
Achilles (DDx on pg. 293)		
Achilles Tendonitis	• Raise heel • More shoe flexibility • Calf stretching program • Correct any hyperpronation (Table 20-3) • Avoid uphill and speed work	• If severe, must discontinue running for a while; rupture possible, scarring with permanent problems often. • Consider short leg walking cast • Steroid injection contraindicated • Surgery last resort
Calf (DDx on pg. 293)		
Calf Strain	• Raise heel • More shoe flexibility • Calf stretching program • Avoid uphill and speed work	• Rule out DVT, sciatica, claudication. • Running through significant discomfort may increase permanent disability
Shin (DDx on pg. 293)		
Medial Distal Tibial Syndrome ("shinsplints")	• Correct hyperpronation (Table 20-3) • Reduce shock (Table 20-2) • (May be hard to find proper balance between the two)	• Can progress to tibial stress fracture (increased pain, localized tenderness); watch for it

Tibial Stress Fracture	• 3 Common types:
	  a. slow progression from MDTS (above)
	  b. metaphysis
	  c. mid-shaft (esp. likely to complete)
	• Must discontinue running, jumping, prolonged walking until asymptomatic and healed by x-ray
	• Then as activity slowly resumes correct hyperpronation for type a, reduce shock (all types)
		• If severe or mid shaft, consider casting or orthopedic referral

Fibular Stress Fracture	• Discontinue activity until healed • Then slow resumption activity	
Anterior Compartment Syndrome (recurrent)	• Lower heel • Avoid downhills	• May progress to acute compartment syndrome unpredictably; warn patient • Consider referral for elective fasciotomy

Knee (DDx on pg. 293 and Chapter 15)

Patellofemoral Pain ("runners knee")	• Reduce shock (Table 20-2) • Hamstring stretching program • Quadriceps strenghtening (Patient Hand-Out 15-1)	• Reduce hyperpronation if present (*may* help) (Table 20-3) • See Chapter 15 too
Patellar Tendonitis	• Reduce shock (Table 20-2) • Hamstring and quadricep stretching • Compression	• Steroid injection contraindicated • If severe, risk of rupture with continued activity and of chronic irreversible scarring
ITB Tendonitis	• Avoid inclines and hills • Correct leg length discrepancy • Watch for and avoid "scissoring gait" (feet crossing) • Reduce shock (Table 20-2)	• ITB stretching • Steroid injection may be offered

Thigh

Hamstring Strain	• Gentle hamstring stretching program • Decrease stride length • Avoid hills and speed work • Strengthen hamstrings to achieve Q:H strength ratio of 3:2 • Compressive wrap	• Often prolonged problem and permanent limitations • Physical therapy modalities can help • Be sure not torn (weakness of knee flexion)
Adductor Strain	• Stretching program	
Femoral Stress Fracture	• Discontinue activity (*no* weight bearing especially if severe or femoral neck) • Orthopedic referral (STAT if severe or femoral neck)	
Trochanteric Bursitis	• ITB stretching, hip rotator stretching • Steroid injection may be offered (Figure 14-4)	• Avoid inclines • Correct leg length discrepancy

Groin

Groin Strain	• Stretching and strengthening involved muscles • Physical therapy modalities	• Reduce speed, avoid hills • Often difficult problem

PATIENT HANDOUT 20-1

Preventing Running Injuries

Most running injuries are *overuse* injuries, which is to say they occur because the stress on an area is increased too much, too fast, and it breaks down. Such injuries can be prevented by listening to your body . . increasing your mileage and speed slowly enough for your bones and muscles to adapt, and slowing down your training if one body part or another becomes especially sore. An alternate day schedule, or at least alternating heavy with light days, is best.

Also be sure to run only in good quality, well fitting running shoes, and to replace them after a few hundred miles; by then they have often lost a lot of their cushioning ability even if they do not show much wear.

Be aware that hills, hard surfaces, inclined surfaces and higher speed all put extra stress on your body. And be sure to warm up and stretch properly beforehand.

Good running!

PATIENT HANDOUT 20-2

Stretching

All stretching should be done only after muscles are warmed up. Read that again: stretching is not in itself warm-up, and stretching cold, inflexible muscles is more likely to lead to trouble than anything else. So first move around for a while, jog half speed, do jumping jacks gently or whatever; and *then* stretch.

Each muscle group should be stretched in turn as shown; stretch *gently* and *steadily,* almost to the point of pain but not quite, and hold each stretch for thirty to sixty seconds. *Do not bounce.*

Stretching again after running can help to increase flexibility and prevent muscle soreness.

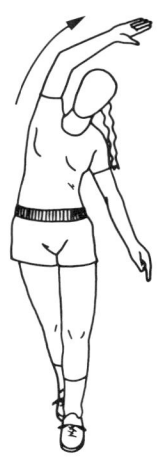

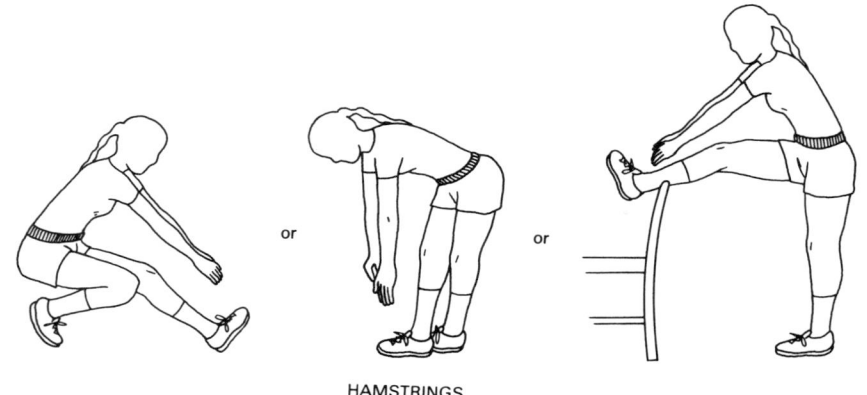

HAMSTRINGS

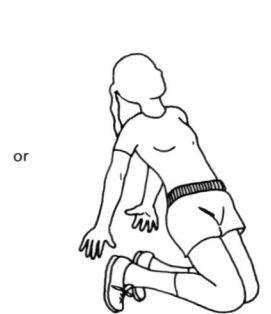

QUADRICEPS

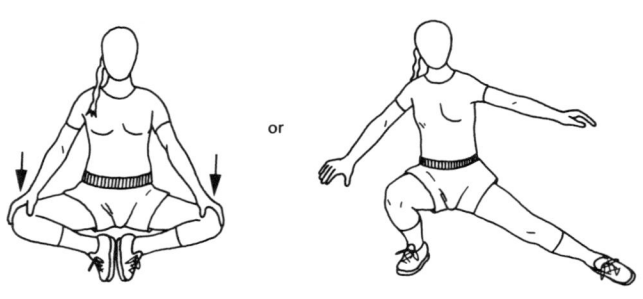

GROIN AND ADDUCTORS

CALF

ACHILLES

PATIENT HANDOUT 20-3

Healing Your Running Injury

Cut back activity enough so that things are getting better rather than worse. No one can tell you how much; you'll have to "listen to you body" and find out yourself. If your injury is severe or persistent a period of total rest is advisable. For some injuries you may *have* to lay off completely; your doctor will tell you. (If you have to stop running you may be able to substitute another activity for a while.)

As healing takes place you may slowly re-increase your training. Again, listen to your body and don't go too fast too soon. Running only on alternate days is best, or at least alternate your heavy days with light ones.

Put an ice pack over the sore area for about twenty minutes after running, and several times a day if needed.

Unless there is some reason you shouldn't, take a couple of aspirin one half hour before running; and if necessary four times a day. (Check with your doctor if it's okay; you may be given a prescription for another antiinflammatory medicine, and you shouldn't take both that *and* aspirin.)

Make sure you warm up and stretch properly before running; pay special attention to any particular exercise your doctor has emphasized.

- ☐ Decrease transmitted shock up your leg; see attached handout.
- ☐ Decrease pronation; see attached handout.
- ☐ Other instructions:

Decreasing Shock

Every time your foot hits the ground running a force of two to three times your body weight is sent up your leg. Decreasing this force will help prevent injury.

1. Avoid downhills as much as possible.
2. Run on terrain that is as soft as possible.
3. Slow down a bit.
4. Choose shoes that have adequate cushioning, and replace them every few hundred miles. (The cushioning can be lost before the shoe looks very worn out).
5. Add a visco elastic cushion to your shoe.
6. For you serious runners, a good coach may be able to make some suggestions about your running style that will help.

PATIENT HANDOUT 20-5

Controlling Pronation

Runners who have especially flexible feet (flat feet) often wind up with certain types of injuries because of it. It this is your problem, you can help control it by:

1. Staying off of inclined (banked) surfaces.
2. Choosing shoes that:
    - have a snug stiff heel counter (and if it is reinforced it will last longer)
    - are built on a curved last
    - have good arch support
    - have a denser sole toward the arch side of the shoe
3. Adding a medial heel wedge or arch support - ask your doctor.
4. If your problem is bothersome enough and these simple measures don't help, you may have to have a customfitted *orthotic* made for you. These can be expensive but are often well worth it. Again, ask your doctor.

21

Muscle Problems

Muscle Trauma 307
    Muscle Pull 307
    Contusion (Bruise) 308
    Myositis Ossificans 309
Diffuse Muscle Pain 310
    Evaluation 310
    Fibrositis 310
    Polymyalgia Rheumatica 311
Muscle Cramps 311
Compartment Syndromes 312

The differential diagnosis of *diffuse muscular weakness,* sometimes associated with pain, obviously includes various diseases of muscle, the motor end plate, and the peripheral and central nervous system, and is beyond the scope of this chapter. The reader is referred to textbooks of internal medicine or neurology for discussion of the possibilities. Keep in mind that unexplained *muscle masses* can be *malignant.*

MUSCLE TRAUMA

Muscle Pull (Strain) (*Mild Ones Extremely Common, Severe Ones Uncommon*)

Diagnosis

    The history is of a single excessive forceful stretch or of repeated mild excessive stretch of the involved muscle.

There will be local pain and tenderness in the involved area, and the involved muscle may be in spasm.

Ecchymosis and/or impartial strength of the involved muscle indicate at least some tearing of the muscle (this is called a *second degree strain*). If there is a palpable defect in the muscle, or if it is noted to "bunch up" toward either its origin or insertion, a complete tear (*third degree strain*) is present, and there will be significant loss of function.

Check distal neurovascular status and keep in mind the possibility of a compartment syndrome.

Pathophysiology

Acute strains are usually the result of a sudden streching force for which the muscle is not prepared, or which is greater than its ability to resist.

Chronic strains are a result of the muscle having insufficient flexibility for the stretch being repetitively applied. An imbalance of strength with its counteracting muscle can also be a contributing factor.

As noted above, first degree strains consist of stretched muscle, second degree strains of partial tears, and third degree strains of complete tears.

Treatment

An acute first degree or mild second degree strain should be treated with *ice, compression,* and *elevation,* and it should be *rested* until syptoms diminish.

Following such an acute strain, a program of nonballistic *stretching exercises* should be carried out to help prevent recurrence.

In chronic strains, exercises to stretch the involved muscle are most important. *Strengthening* the muscle in relation to its counteracting muscle is often helpful as well. (For example, repetitively strained hamstrings may be too weak in comparison to the quadriceps.)

Severe second degree and all third degree strains should be referred to an orthopedist, since often *surgical repair* is warranted, and the decision whether or not to operate is often a complex one. If repair is to be done, it must be performed immediately.

Contusion (Bruise) (*Extremely Common*)

Diagnosis

After direct trauma, ecchymosis develops. If bleeding is deep into a large muscle mass, though, discoloration may not be evident, at least early. In this case the history and the complaint of deep pain increased with motion, along with a negative X-ray, is sufficient to make the diagnosis.

If bruising is repeated, excessively easy or unexplained, the possibility of a clotting problem should be considered and investigated. In children, keep in mind the possibility of child abuse.

MUSCLE TRAUMA

Pathopsysiology

The hematoma can be subcutaneous or within the body of the muscle. Less commonly it is subperiostial (a "bone bruise"; these remain tender for long times.)

The hematoma eventually is "organized" by the development of granulation tissue and fibrous scar tissue.

Treatment

Acutely, *ice, compression,* and *elevation* should be used.

Very large bruises that are not clearly superficial should be *rested for several weeks* to help prevent the development of MYOSITIS OSSIFICANS. *Massive* hematomas may need to be *surgically drained.*

Myositis Ossificans* *(Uncommon)*

Diagnosis

Pain and tenderness persist in the area of a large hematoma, most commonly in the thigh or upper arm.

Calcification can be evident on X-ray a few weeks after injury. Ossification is usually complete in several months. If large, the bony mass is palpable.

This condition can be confused with bone tumors; obviously referral for biopsy is necessary if there is any doubt.

Pathophysiology

An organizing hematoma becomes calcified and is then invaded by osteoblasts which form new bone. The precise trigger and mechanism are unknown.

Treatment

The involved muscle should be rested (*not* immobilized) for several months, and then gentle strengthening exercises begun.

Otherwise only *symptomatic measures* are necessary unless significant disability persists, as is often the case if the mass is near a joint. Then the mass can be surgically removed once maturation is complete (generally about a year postinjury).

---

*The condition discussed here is classified as *myositis ossificans circumscripta* or *myositis ossificans traumatica,* becuase it consists of the ossification of an organizing hematoma. There is a quite distinct disease of unknown etiology called *myositis ossificans progressive* or *generalized myositis ossificans,* which usually strikes young children or young adults, is often associated with congenital deformities, and progresses in relentless fashion to involve the entire body. No effective treatment is known.

21. MUSCLE PROBLEMS

DIFFUSE MUSCLE PAIN

Evaluation

Obviously, by far the most common cause of acute diffuse myalgia is acute systemic infection, usually viral. Keep in mind that acute diffuse muscle inflammation, whether from viral infection or overuse, can sometimes be severe and can lead to myoglobinuria and acute renal failure.

Long-term chronic diffuse muscle aching is usually related to deconditioning, degenerative arthritis with secondary muscular effects, or to psychological factors. Chronic muscle achiness with stiffness, fatigue and multiple trigger points may indicate FIBROSITIS, covered below.

Subacute diffuse myalgia can be the presenting symptom of rheumatoid arthritis, other rheumatic diseases, or polymyositis. Other evidence for these conditions should be sought. But consideration must always be given to the possibility of POLYMYALGIA RHEUMATICA, discussed below.

Fibrositis (fibromyalgia, myofascial pain syndrome, non-articular rheumatism) (*common*)

Diagnosis

This is a chronic condition in which the chief complaint is diffuse muscle aching, especially proximal, which fluctuates from day to day.

Stiffness and fatigue are usually associated. There is often poor sleep and often depression.

Symptoms are usually fairly clearly worse with stress, with cold, damp weather and after excessive exercise; and better with warmth, gentle exercise, massage and relief of stress.

Examination is usually normal except for the presence of multiple small, firm, very tender areas of muscle called *trigger points*. Strength and joint examination is normal.

Rheumatologic and muscle enzyme tests are normal.

It must be differentiated from PMR, polymyositis, rheumatoid diseases and hypothyroidism. "Secondary fibrositis" can be associated with rheumatoid disease.

Pathophysiology

Although obviously a common and fairly well-defined clinical syndrome, the pathology of the condition has never been elucidated.

Depression or sleep disturbance may play a large role through mechanisms as yet undefined.

MUSCLE CRAMPS

Treatment

Education of the patient about the chronic but nonprogressive nature of his or her affliction is often reassuring.

Various *stress management* and *relaxation techniques* can often be very helpful; consider referral for such, and/or advising regular *gentle* aerobic *exercise* such as swimming in a heated pool. Known precipitating factors should be avoided as much as possible.

Antidepressant medication, starting with low doses, often helps relieve associated sleep disturbances as well as fatigue.

Aspirin or other *NSAI's* can be prescribed. Occasional *injection* of especially painful trigger points should be considered (see Figure 4-2).

Consider referral to a physical therapist for stretching, massage, or other modalities.

Polymyalgia Rheumatica (*Rare*)

Clinical Features and Diagnosis

Usually occurs in patients above the age of 50.

The patient complains of diffuse aching pain, most commonly in the neck and shoulders, and sometimes in the hips, thighs and arms as well. The pain is worse after motion, and there is morning stiffness.

The sedimentation rate is significantly elevated. The RF and ANA tests are negative.

A large number of these patients have associated temporal arteritis. This is of course potentially a much more serious condition. Opinion differs as to whether routinely to perform a temporal arterial biopsy on all patients with PMR before instituting treatment, as arteritis can be spotty, leading to a false-negative result.

Pathophysiology

Unknown.

Treatment

Oral steroids are used. Because of the occasional difficulty in excluding temporal arteritis, and because the latter condition requires much higher dosages of these dangerous medications, *referral* to a rheumatologist is recommended.

MUSCLE CRAMPS (*Very Common*)

Diagnosis

The patient reports sudden contractions of a muscle with the development of a muscular knot, lasting up to several minutes.

The calf and foot are the most commonly involved areas, but the cramps can occur anywhere.

They are most commonly due to activity excessive for the condition of the muscle, and occur either during the activity or hours later.

They also occur frequently in pregnant women secondary to lack of sufficient calcium intake.

Pulses and capillary filling must be checked to rule out vascular compromise as the cause of the muscular irritability, especially in older patients.

Tests for serum, sodium, potassium, calcium, magnesium, phosphorus, and sugar and a CBC should be performed in patients with frequent recurrence, since diabetes, polycythemia, or abnormalities of electrolytes can be responsible.

Treatment

The patient should be instructed that when a cramp develops he should keep the muscle *stretched* until it abates. (In the case of the calf, by dorsiflexing the foot with the hand or by pushing against something with the forefoot).

Specific etiologies discovered in the evaluation discussed above should be dealt with.

If none of these are found, the patient should perform *stretching exercises* each morning. Teach or check the calf stretching exercises in Patient Handout 20-2.

Quinine sulfate (5 grains at bedtime) has been used with some success in the prevention of nocturnal cramps.

COMPARTMENT SYNDROMES (*Uncommon*)

Diagnosis

Although seen perhaps more commonly after severe trauma, vascular occlusion or orthopedic surgery, compartment syndromes can also occur as a result of exercise.

The patient will complain of pain in the involved compartment and sometimes will complain of a neurologic deficit as well. By far the most common exercise-induced compartment syndrome involves the anterior compartment, which can give numbness in the first web space and weakness of the toe extensors. (*Note:* A decrease in two-point discrimination is the earliest, most subtle sign of neurologic compromise.)

Tenderness and sometimes palpable tenseness will be found over the affected compartment, and pain will result from *passive* stretch of the muscles there. (In the case of the anterior compartment, passive toe and foot plantar flexion will hurt.)

COMPARTMENT SYNDROMES

This problem can be present in an acute and progressive way (in which case the patient must be seen by an orthopedist *STAT*), or, more commonly when running-induced, as a mild, self-resolving, recurrent complaint after exercise, often with no findings at all when seen by the doctor.

Pathophysiology

In exercise-related cases, increased capillary permeability leads to swelling and thus increased pressure in the closed space. This can progress to ischemia and thus set up a vicious cycle.

Predisposition to the problem depends on anatomy, i.e., the relative size of the involved compartment and its musculature.

Swelling can obviously also be due to trauma or surgery.

The increased pressure then can lead to peripheral nerve compression and destruction if it is not resolved spontaneously or surgically.

Treatment

Acute or progressive cases or those in which neurologic findings are present at the time of examination *must* be evaluated by an orthopedist within no more than a few hours.

Patients who give a history suggestive of mild recurrent compartment syndrome should be aggressively treated to remove excessive strain on the muscle groups involved, as discussed in the appropriate section. Referral for elective fasciotomy should be strongly considered in all but the mildest cases, since no guarantee can be given that the problem will not advance to the acute state at any time.

22

Degenerative Arthritis and Osteoarthritis

Clinical Features 314
    Degenerative Arthritis. 314
    Osteoarthritis. 315
Diagnosis. 315
Pathophysiology 316
Treatment 316

These terms are used interchangeably to describe a single pathologic entity, but they can be considered to represent two quite distinct (though often coexistent) clinical symptoms.

CLINICAL FEATURES

Degenerative Arthritis

    This is the wearing-out of one or more joints. It is a ubiquitous and (so far) unavoidable consequence of aging.
    Joints which have been overused or placed under more strain wear out sooner.
    Joints which have been directly traumatized or had their mechanics altered by trauma, deformity or previous disease (such as rheumatoid arthritis or aseptic necrosis) likewise will degenerate before other joints in that individual.

DIAGNOSIS

The hip, knee, first MCP and first MTP joints, and the facet joints of the spine are most commonly involved, but any and all joints can become symptomatic.

Patients complain of aching pain in the joints, worse after usage or weight bearing; there is some stiffness after rest but this is not nearly as dramatic as in patients with rheumatoid arthritis.

Note that deterioration with age and cumulative microtrauma is not limited to joints; it occurs in tendons, muscles, etc. as well. Perhaps DEGENERATIVE MUSCULOSKELETAL DISEASE would be a better term.

A severe progressive form occurs in joints which lack sensory innervation because of neurologic disease. These are called *Charcot Joints*. (They are not usually painful.)

Osteoarthritis

This condition mainly affects small joints, most commonly the DIP's, where bony hypertrophy at the sides of the joints produces Heberden's nodes.

Less commonly, the PIP's are affected, giving rise to Bouchard's nodes.

This is primarily a disease of women; it can begin as early as the fourth decade.

There is a definite familial predisposition.

Pain and stiffness are usually much milder than in rheumatoid arthritis.

An uncommon form called *erosive osteoarthritis* can occur, with a single small joint or several becoming acutely inflamed and painful for several weeks.

DIAGNOSIS*

There are no systemic manifestations.

On examination, there may be joint line tenderness, pain on (and sometimes limitation of) range of motion, and occasionally crepitus. In joints in which the condition is more advanced, the bony deformity will be obvious.

Redness and warmth are lacking (except in erosive osteoarthritis).

X-ray may be normal early, when the disease is confined to the articular cartilage; but later it becomes characteristic, showing joint space narrowing, bony hypertrophy (especially at the joint margins), and sometimes periarticular sclerosis without osteoporosis. A word of caution: these radiographic changes are ubiquitous in older patients and previously injured joints, and their presence does not rule out another cause for the specific complaint under investigation.

Blood test, including the ESR, are normal.

---

*The material from this point through page 339 applies to both the degenerative and osteoarthritic forms of the disease.

Synovial fluid analysis reveals findings typical of noninflammatory conditions, with white cell counts in the 500 to 5000 range and consisting mostly of mononuclear cells.

PATHOPHYSIOLOGY

The first change found is in the articular cartilage, which becomes yellowed and focally softened and roughened. These changes then spread to involve the entire joint surface, which becomes progressively more frayed and cracked as time goes on.

Along with these changes, there occurs remodeling of the periarticular bone, with sclerosis of the bone underlying the articular cartilage and development of spurs of bone (osteophytes) at the joint margins as well as at the sites of attachment of tendons and ligaments. These changes may be a result of the normal remodeling process that occurs in bone with changes in applied stress, in this case the changes in applied stress being secondary to alterations in the configuration of the joint.

Synovial inflammation, the level of which is quite variable, can result from the stresses of altered joint mechanics, or from direct irritation of the synovial membrane by tiny fragments of articular cartilage released into the joint space.

Eventually all the articular cartilage can be worn away, leaving a joint consisting of bone on bone. Large fragments of cartilage and bone can become loose bodies within the joint, leading to locking and collapsing.

Various progressive histologic and biochemical changes have been observed in involved cartilage, but the precise mechanism that leads to the gross pathologic changes that are seen remains a mystery.

A large part of the pain may actually arise from fatigue and tension in muscles surrounding the joint, rather than from joint inflammation itself.

TREATMENT

It is often important to *reassure* the patient that he does not have a progressive systemic disease, lest the diagnosis of arthritis lead to visions of wheelchairs and such (Patient Handout 22-1).

A proper *balance of rest and use* of the joint is important for symptomatic relief, and probably to help slow progression of the disease as well. The specific amount of each is best determined simply by the patient's symptoms. Too much use or not enough will lead to worsening of discomfort and stiffness. It must be emphasized to the patient that if his joints are hurting, they are telling him something; they need to rest.

TREATMENT

Weight loss is very important when obese patients have involvement of the weight bearing joints. Better *shoe cushioning* may help as well.

Local *heat* has an analgesic effect (see the discussion in Chapter 3).

Salicylates and other *analgesic/antiinflammatory* medications are mainstays of therapy. Again, see Chapter 3 for more details.

Specific measures to avoid stress to involved joints (such as use of a cane in degenerative arthritis of the hip or a cervical collar when the neck is symptomatic) are discussed in the sections on degenerative arthritis in the chapters on the individual joints.

Steroid injection into selected troublesome joints can give temporary but prolonged relief. However, it is quite possible that this is at a cost of accelerating cartilagenous destruction and thus the underlying disease. An individual joint should probably not be injected more than two or three times a year. The procedure is discussed further in Chapter 3, and techniques for entering specific joints will be found in the corresponding chapters.

In moderately severe cases, referral to a *physical therapist* is advisable for training in range of motion and strengthening *exercises* to help maintain joint mobility and stability, and to restore proper joint mechanics as much as possible. The therapist can also be helpful in the provision of warming modalities such as ultrasound and paraffin baths.

Referral for *surgery* should be considered in severe disease which does not respond to the conservative measures described above. Several types are performed, depending on the individual case: joint debridement, joint fusion, osteotomy (where the weight distribution on the joint surfaces are altered; most common in the knee), or total joint replacement with a prosthesis. The later is still most commonly used in the hip and knee, but the development of other artificial joints is proceeding rapidly.

PATIENT HANDOUT 22-1

Degenerative Arthritis (Osteoarthritis)

Your doctor has determined that your joint pain is due to DEGENERATIVE ARTHRITIS (sometimes called DEGENERATIVE JOINT DISEASE of OSTEO-ARTHRITIS). You should be relieved to some extent: it means that you *don't* have a crippling disease like rheumatoid arthritis.

WHAT IS IT?

The surfaces of joints wear out. If you live long enough, all of your joints will wear out; so will everyone else's. But some people's joints wear out sooner than others, and doctors aren't sure why. They *do* know that joints that have been injured, or that have excessive strain placed on them, wear out much faster than others. (Examples would be an ankle that has been sprained severely, or the hips in someone that is overweight, or the base of the thumb in a mailman who pinches piles of letters all day.) In some people, the joints of the fingers begin wearing out for unknown reasons at a relatively young age. This wearing out of the joints is what is called DEGENERATIVE ARTHRITIS.

WHY DOES IT HURT?

Because the joint surface is rough instead of smooth, it hurts to move, at least after a while. A joint will feel stiff because the muscles around it tighten up.

WHAT CAN I DO ABOUT IT?

A lot of things!

1. Find a good balance of resting the joint and using it. The way to do that is to find by trial and error the amount of use that leads to the least pain during the day as a whole.

PATIENT HANDOUT

2. Aspirin not only helps take the pain away, but it actually reduces joint inflammation. If your joints only hurt occasionally, just take it when you need it; if they hurt a lot of the time, take a couple of aspirin four times a day on a regular basis. Never take it on an empty stomach. If it upsets your stomach anyway, try using a preparation of aspirin that has antacid mixed in. If it still leads to problems, of if you can't take aspirin because you are allergic to it, have a history of ulcers, or are on blood thinners, or if it doesn't seem to be doing the job, let your doctor know. There are other medications he or she can prescribe for you.
3. Heat or cold are often helpful. Your doctor may give you some information on how to use them best.
4. If you are overweight and your problem is in your back, hips, knees, ankles, or feet, losing weight is one of the best things you can do for your joints.
5. If all these things don't give you sufficient relief, let your doctor know. He might prescribe splints, exercises, physical therapy, or recommend an injection. Rarely, surgery may be considered.

23

The Rheumatoid Diseases

Rheumatoid Arthritis. 321
Systemic Lupus Erthematosis 322
Psoriatic Arthritis 322
Enteropathic Arthritis 323
Reiter's Syndrome. 323
Other Arthritides 324

Various patterns of joint involvement are more or less typical for certain diseases, but these patterns cannot be relied upon too heavily in arriving at a diagnosis. The evaluation of polyarthritis always requires a complete history and physical examination, since the cause is often a systemic illness, and findings in other organ systems can be crucial to determining the proper diagnosis. Laboratory tests are often also needed (see Chapter 2 for a discussion), and X-rays (especially of hands and spine) can be helpful. This chapter is by no means intended to be more than an outline of the more common arthritides; the reader is referred to a textbook of internal medicine or rheumatology for further information of these conditions.

Aside from the conditions reviewed in this chapter, GONOCOCCAL ARTHRITIS and VIRAL ARTHRITIS should be considered in acute oligoarthritis (Chapter 25); and SEPTIC ARTHRITIS (Chapter 25) or crystalline disease (Chapter 24) can sometimes affect more than one joint at a time. DEGENERATIVE ARTHRITIS and OSTEOARTHRITIS are covered in Chapter 22. ANKYLOSING SPONDYLITIS is discussed in Chapter 11, and JUVENILE RHEUMATOID ARTHRITIS and RHEUMATIC FEVER are covered in Chapter 19.

Rheumatoid Arthritis *(Common)*

Clinical Features

The usual onset is with involvement of several small or medium-sized joints, in a fairly symmetrical distribution.

The disease can, however, begin with a single inflamed joint.

There often are prodromal or associated systemic complaints such as stiffness and fatigue.

Morning stiffness is such a characteristic symptom that its duration can be used as an indicator of the level of disease activity.

The course of the disease usually is marked by spontaneous remissions and exacerbations. Severity of the disease varies tremendously, from patients with mild intermittent symptoms to those (fortunately less common) who are eventually crippled.

Some patients have rheumatoid nodules, which are subcutaneous and usually found at pressure points such as the olecranon.

Sjogren's syndrome (keratoconjunctivitis sicca, dry mouth and eyes) is seen in some patients; less common associated conditions include uveitis, episcleritis, vasculitis, pulmonary fibrosis, and polymyositis. *Felty's syndrome* is a rare combination of RA, splenomegaly, and neutropenia.

Females outnumber males three to one.

Once involved, a joint tends to remain involved as the disease spreads to other areas.

Examination early in the involvement of a joint will reveal warmth and tenderness; later on, synovial thickening will be palpable; and eventually loss of range of motion and various deformities will be evident.

The ESR is elevated and can be used to follow the level of disease activity.

The rheumatoid factor is positive in most patients with the disease but can sometimes be negative, especially early in the course.

A fair percentage of patients will have a positive ANA test.

Synovial fluid of involved joints will be typical for inflammatory conditions, with WBC counts in the 5,000 to 50,000 range and consisting mainly of PMN's.

X-rays of involved joints show soft tissue swelling early, and then joint-space narrowing and progressive periarticular osteoporosis and bony erosions.

Pathophysiology

Although autoimmune factors no doubt play a part, the full picture of the etiology of this remains a mystery.

The synovium is inflamed and edematous. It becomes hypertrophic and spreads to cover the articular cartilage, destroying it and eventually the bone underlying it as well. This can then lead to adhesions in, and ankylosis of, the joint, as the supporting capsule is weakened by the inflammation.

Treatment

Local *heat* often gives symptomatic relief. For the hands, a paraffin bath is an easy way to provide this.

Range of motion and strengthening *exercises* are important in the maintenance of joint mobility and stability and thus the prevention of contractures and other deformities.

Referral to a physical therapist for training in and provision of the above modalities is indicated except in mildly affected individuals.

Aspirin or other *nonsteroidal antiinflammatories* are the first line of treatment. See Chapter 3.

Injection of steroid into severely affected joints can give significant relief for months. See the discussion in Chapter 3. Techniques are shown in the chapters on the individual joints.

In severe, progressive or unresponsive cases, *disease-suppressant medication* should be given. These are dangerous drugs, and the patient should be handled by a physician familiar with their use. Medicines now being used include gold salts, chloroquin, penicillamine, corticosteroids, and various cytotoxic drugs.

Surgery, either synovectomy and/or procedures to correct deformities and contractures, is often necessary in severely affected joints.

Systemic Lupus Erythematosis *(Rare)*

This is a systemic autoimmune disease of unknown etiology. Fleeting joint pains as well as frank joint inflammation are the most common complaints. The arthritis rarely leads to any significant joint destruction or deformity. The disease also affects (in varying percentages of patients) serosal surfaces, the lymphoid system, the skin, the kidneys, and the CNS. Involvement of the latter two organ systems is often life-threatening. Corticosteroids are the first line of treatment. The reader is referred to texts of internal medicine or rheumatology for further information.

Psoriatic Arthritis *(Uncommon)*

Clinical Features

Most commonly there is an asymmetric polyarthritis involving a few small joints. The DIP joints are always very commonly involved; nails adjacent to involved DIP joints are almost always pitted.

Although the activity of the joint disease tends to parallel activity of skin disease, either manifestation can appear first, so the absence of psoriasis of the skin does not rule out psoriatic arthritis.

About one-fourth of patients will have spondylitis, which is radiographically different from that seen in ankylosing spondylitis.

The sedimentation rate can be elevated; the rheumatoid factor is negative; uric acid is often elevated is psoriasis, so this should not lead to an erroneous diagnosis of gout. (Of course, sometimes two diseases are present.) The HLA-B27 Antigen is positive in 70% of the cases in which spondylitis is a factor.

A destructive, progressive form, called *arthritis mutilans,* rarely occurs.

X-ray changes are fairly characteristic, showing destruction of scattered small joints and occasionally the "pencil in cup" deformity at involved DIP's as well as fluffy periostitis.

Pathophysiology

The pathology of the synovial disease is similar to that of rheumatoid arthritis.

Treatment

Antiinflammatory medications and *physical modalities* are used as in rheumatoid arthritis.

In severe destructive disease, the patient should be referred for consideration of use of *cytotoxic agents* such as methotrexate.

Enteropathic Arthritis *(Rare)*

Clinical Features

About one-fifth of patients with ulcerative colitis and a smaller fraction of patients with regional enteritis have an associated arthritis, which takes one of two forms.

The more common type finds several large joints, more commonly of the lower extremity, becoming inflamed in succession. These attacks tend to resolve in weeks or months, leaving little or no residual damage. Activity of the joint disease tends to parallel that of the gastrointestinal disorder.

A smaller percentage of patients develop a spondylitis indistinguishable from that of ankylosing spondylitis.

Treatment

The peripheral arthritis is treated with *rest* and *antiinflammatory medication,* and by *treating the underlying GI disease.*

The spondylitis is managed the same as ankylosing spondylitis (covered in Chapter 11).

Reiter's Syndrome *(Uncommon)*

Clinical Features

The classical triad consists of arthritis, urethritis, and conjunctivitis. This is a disease almost exclusively of young adult males.

The arthritis is usually asymmetric, and can involve a few or many joints, both large and small. Achilles tendonitis and plantar fascia inflammation often occur. Occasionally spondylitis is present. The arthritis usually persists for several weeks to several months and clears spontaneously. The spondylitis can be more persistent.

The urethritis must be distinguished by gram stain and culture from gonorrhea, which of course can also give an acute oligoarthritis.

The conjunctivitis is usually mild and transient; keratitis is rare.

Mucocutaneous lesions are common. *Keratoderma blennorrhagica* consists of painless pustules on the sole and plam, and is indistinguishable from pustular psoriasis. Occasionally painless shallow ulcerations of the buccal mucosa are found.

Recurrences are not uncommon.

Pathophysiology

It is likely that somehow the infecting organism of nonspecific urethritis (ureaplasma or chalmydia) lead to the other manifestations of the syndrome, but the mechanism by which this occurs remains unknown. The fact that the same syndrome sometimes occurs in association with shigella infections complicates the picture.

Treatment

Rest and *antiinflammatory* drugs should be used to suppress symptoms until the arthritis resolves.

Some physicians advocate a ten-day course of an antibiotic (tetracycline, erthyromycin or trimethoprim/sulfa) even if the urethritis is no longer (or never was) present.

If any suspicion remains that the patient has gonococcal arthritis, treatment for this disease should be given as well. See Chapter 25.

Other Arthritides

An oligoarthritis can occasionally occur in *bacterial endocarditis*. Joint inflammation can sometimes precede the onset of the other features of *serum sickness*. Patients with *sarcoidosis* may have a migratory self-limited polyarthritis, or more likely a chronic arthritis related to granuloma deposition around the joint. *Behĉet's syndrome* is a rare disease probably of viral origin, in which an intermittent oligoarthritis is a very common finding. (Other components of the syndrome are aphthous stomatitis, genital ulceration, iritis, and CNS involvement.) Attacks of *Familial Mediterranean Fever* can include synovitis of large joints. Various inborn errors of metabolism such as *Gaucher's disease* and *alcaptonuria* and inherited connective tissue disorders such as the *Ehlers-Danlos syndrome* can lead to joint pathology. Arthritis may rarely be due to local or remote *neoplastic disease,* or to *metabolic disease.* Further discussion of these conditions is beyond the scope of this book.

24

The Crystalline Diseases

Gout...325
Pseudogout 326

Gout *(Common)*

Clinical Features and Diagnosis

The condition is much more common in males.

The disease presents with one or more acute attacks. The first MTP joint is involved in fully one-half of first attacks (where it is called *podagra*); other commonly affected areas are the heel, instep, ankle, and knee. The affected joint becomes suddenly red, warm, swollen and extremely painful. Untreated attacks last for days to weeks.

Attacks can be set off by fasting, alcohol binges, overuse, or the administration of diuretic or hypouricemic drugs; most often, however, no specific precipitant is noted.

The serum uric acid level is usually elevated, but the only way to make the diagnosis with absolute certainty is to tap the joint and find negatively birefringent uric acid crystals.

If hyperuricemia is untreated, over a decade or more the patient may develop *chronic gouty arthritis,* in which there are uric acid deposits in articular cartilage and other soft tissue structures. *Tophi* also occur. These are urate deposits with surrounding inflammation, most often found in the pinna, tendons and superficial bursae.

A CBC and renal function tests should be performed.

About one-third of patients with gout will suffer at some time from urinary tract stones.

Pathophysiology

Uric acid is a normal by-product of nucleic acid metabolism. Hyperuricemia is due to uric acid overproduction or underexcretion, often of unknown cause. (The definition of abnormally high levels of uric acid in the blood must be an arbitrary one, since their distribution in the population follows a bell-shaped curve). Sometimes hyperuricemia is secondary to the use of thiazide diuretics or other drugs. Obesity and high-protien diets may be contributory. Infrequently the abnormality is a result of a known specific enzymatic defect. It can rarely be due to rapid purine turnover seen in the myeloproliferative disorders or chronic hemolytic anemia, or to decreased excretion in chronic renal disease.

In patients with hyperuricemia, urate exists as a supersaturated solution in most body fluids. Rapid fluctuations in level or possibly other factors then lead to the precipitation of crystals in a joint, i.e., an acute attack.

After many years, deposition of uric acid in various tissues can lead to chronic gouty arthritis and tophi.

Treatment

The acute attack should be treated with *elevation* and potent *antiinflammatory mediciation*. *Colchicine* has been the drug of choice for a long time, but in full therapeutic doses has more GI side effects than drugs such as indomethicin.

If hyperuricemia is thought to be drug-induced, consideration should be given to discontinuance of the suspected medication.

The higher the uric acid level, the more likely the patient will suffer recurrent attacks. With levels over 9 mg % they are almost certain to occur, and the development of chronic gouty arthritis in later years becomes a real possibility. With these facts in mind a decision must be made about whether to use *medication to lower acid levels*. If so, see Table 24-1. Levels should be followed periodically. Colchicine at a dose of 0.5 mg BID should be prescribed along with the hypouricemic drug for several months, to help prevent the acute attacks which can be precipitated by dropping uric acid levels.

Patients with *asymptomatic hyperuricemia* need not be treated unless levels are very high.

Pseudogout *(Uncommon)*

Clinical Features and Diagnosis

There is acute inflammation of a single joint, most commonly the knee, in an older patient.

Table 24-1. Hypouricemic Drugs[a]

Drug	Adult dose	Comment
Allopurinol (XYLOPRIM, Burroughs Wellcome; LOPURIN, Boots; also available generically) 100 mg, 300 mg	100 to 600 mg/day (in divided doses if more than 300 mg/day)	Decreases uric acid production by inhibiting xanthine oxidase Decreases uric acid in urine, so useful in stone formers Not affected by salicylates Rare severe hypersensitivity reactions can occur; rash can be first sign
Probenecid (BENEMID, Merck, Sharp and Dohme; also available generically) 500 mg	250 mg BID for first week, then 500 mg BID (can push to 1 g BID if needed)	Salicylates interfere with action Increases uric acid excretion, so not to be used in stone formers; pushing fluids wise in all patients GI upset not uncommon Available in combination tablet with colchicine (COLBENEMID, MDS; or generic) for first few months of therapy
Sulinpyrazone (ANTURANE, Ciba) 100 mg, 200 mg	100 mg BID to 200 mg BID (can push to 400 mg BID if needed)	Salicylates interfere with action GI upset not uncommon. Contraindicated with ulcer. Give with food or antacid Increases uric acid excretion, so same problems as probenecid

[a] See the package insert or Physician's Desk Reference for complete information before prescribing any drug.

The episode resolves in a week or two. Rarely, more than one joint can be involved.

Examination of the synovial fluid reveals short positively birefringent crystals (calcium pyrophosphate).

This X-ray will show linear calcification in articular and meniscal cartilage. (This is called *chondrocalcinosis*. While this is an invariable finding in pseudogout, it can also be seen in asymptomatic knees in older patients, and can be related to other conditions: hyperparathyroidism, gout, hemochromatosis, diabetes mellitus, Wilson's disease and acromegaly.)

Pathophysiology

Possibly the acute attack is precipitated by the release of some calcium crystals from cartilage into the joint space, where they initiate an inflammatory response.

Treatment

Elevation, limited-weight-bearing, and antiinflammatory medications are used.

25

Infection

Joint Infection 329
    Septic Arthritis 329
    Gonococcal Arthritis 330
    Viral Arthritis 331
    Tuberculous, Mycotic, and Syphilitic Arthritis 331
Soft Tissue Infection 331
    Infected Superficial Bursae 331
    Septic Tenosynovitis 332
    Infections of the Various Spaces of the Hand 332
    Cellulitis 332
    Lymphangiitis 333
Bone Infection 333

JOINT INFECTION

Septic Arthritis *(Uncommon)*

(*Note:* Gonococcal arthritis is discussed separately below.)

Diagnosis

The patient will present with an acutely inflamed joint, usually tender, red, and warm. But sometimes these signs are not dramatic, so all acutely and atraumatically inflamed joints must be tapped to rule out this possibility. Rarely, more than one joint will be involved.

Fever and chills may or may not be present.

The peripheral WBC count will usually be elevated, but not always.

Synovial fluid will show a high WBC count consisting mostly of PMN's. Gram stain may or may not reveal the causative organism. Cultures should be obtained for aerobic and anaerobic bacteria, for gonococcus, and if the picture is not clear, for fungi and tubercle bacilli as well.

Pathophysiology

Infection can involve the synovial lining or the joint cavity itself, or (usually) both. It may be secondary to hematogenous spread or to previous injection or aspiration of the joint; it may spread from nearby breaks in the skin. Occasionally it is an extension of a nearby osteomyelitis.

Treatment

Intravenous antibiotic therapy should be started immediately, specific for the organisms found by gram stain or culture. For recommended medications in individual situations and recommended durations of treatment, see internal medicine manuals or texts, or consult with a specialist in infectious disease.

Daily *needle aspiration* and *saline irrigation* of the joint probably speeds resolution. Determination of intraarticular antibiotic levels can be helpful. Intraarticular instillation of antibiotics is not necessary and can be very irritating.

Immobilization of the involved joint for a short period is symptomatically helpful. Physical therapy may be necessary to maintain or restore full range of motion.

If response to medical management is not prompt, with a significant decrease in pain and an afebrile condition being achieved in 24 to 48 hours, or if the hip or shoulder is involved *or,* if *any* joint is thought to be infected with staphylococcus, orthopedic consultation should be obtained.

Gonococcal Arthritis *(Common)*

Diagnosis

There is usually an asymmetric involvement of several large joints (or sometimes tendon sheaths).

Usually there is associated fever and a few small, pustular, hemorrhagic skin lesions.

A majority of patients with this syndrome of disseminated gonococcemia will have no symptoms of genital infection, so level of suspicion must remain high.

Urethral or cervical, rectal, and pharyngeal cultures should be obtained, as well as blood and synovial fluid cultures. A VDRL should also be done.

In the early stages of joint involvement, sometimes only the synovium is infected and no specific findings are present in the joint fluid.

Treatment

Always check for then-current recommended treatment regimens.

Follow-up urethral or cervical, rectal, and pharyngeal cultures should be obtained after treatment. The patient should be instructed to inform his or her partners so that they may be treated, and a report should be sent to public health authorities.

Viral Arthritis *(Common)*

Diagnosis

An acute, usually mild oligo- or polyarthritis can accompany or sometimes precede the other symptoms of systemic viral infections. Most commonly this is seen with hepatitis A or B, with infectious mononucleosis, and with rubella (and sometimes after rubella immunization).

Treatment

The condition is self-limiting, and only symptomatic treatment need be prescribed.

Tuberculous, Mycotic, and Syphilitic Arthritis *(Rare)*

The reader is referred to textbooks of internal medicine or infectious disease for further discussion of these conditions, which usually present as a subacute or chronic monarthritis.

SOFT TISSUE INFECTION

Infected Superficial Bursae *(Common)*

Diagnosis

Usually the olecranon or patellar bursa is involved, with findings of tenderness, redness, and warmth.

The bursa should be tapped to obtain material for gram stain and culture and to decompress it.

Pathophysiology

Infection usually enters through nearby breaks in the skin. Staphylococcus aureus is by far the most common pathogen.

Treatment

An *antibiotic* effective against staph (semisynthetic penicillin or cephalosporin) should be prescribed, unless gram stain or culture reveals a different offending organism.

Hot soaks can be helpful.

The affected part should be immobilized and compression applied.

If the course is prolonged, the patient must be observed for the spread of infection into nearby bone or joints.

Septic Tenosynovitis *(Rare)*

Diagnosis

Most commonly found in the hand or wrist.

There is redness and extreme tenderness along the course of the tendon, and pain on its motion.

Treatment

Surgically opening the sheath is often indicated, so *referral* to a hand surgeon should be made STAT.

Infections of the Various Spaces of the Hand
(*Rare* except for infection in the volar tuft of the distal phalanx, called a *felon*)

These are potentially disastrous, and should be referred to a hand surgeon STAT. (PARONYCHIAS are excepted: see Figure 9-10.) Be even more concerned if infection is due to a human bite or tooth wound, even if fluctuance is not observed.

Cellulitis *(Common)*

Diagnosis

Diffuse inflammation with redness, warmth and tenderness is found over the involved area. BE CAREFUL: Never tap into a normal joint through superficial infection.

Systemic symptoms can occur.

A culture can be obtained by injecting sterile saline under the involved skin and then aspirating.

Pathophysiology

Infection is usually initiated through a break in the skin.

Streptococcus is the most common organism, especially if the border of the infected area is raised and purplish (called *erysipelas*). *Staphylococcus* is often present instead. Other organisms can be involved, especially in children (consider *hemophilus*) and in infants, diabetics and immune-suppressed individuals (consider *gram-negatives*.)

Treatment

If fever is absent, the involved area is small and the patient is an otherwise healthy adult, treatment can be with *elevation*, the application of *heat*, and *oral antibiotics* (use a cephalosporin or a semisynthetic penicillin effective against staphylococcus unless there is reason to suspect another organism or gram stain or culture results indicate another organism).

If resolution does not occur or the condition worsens, or if the patient has underlying disease, is an infant, has systemic symptoms or a large area is involved, admission for intravenous antibiotic therapy is advisable.

If loculated pus develops, it has to be drained.

Lymphangitis *(Common)*

Diagnosis

A red streak is noted going up an extremity, often but not always from a separately recognized site of infection.

Pathophysiology

This infection of the lymph duct is usually caused by hemolytic streptococcus.

Treatment

Guidelines are identical to those described under cellulitis.

BONE INFECTION

OSTEOMYELITIS should be suspected in any case of otherwise unexplained bony pain or tenderness; a bone scan will often be positive long before X-ray changes are apparent. Since biopsy of bone is often required to obtain culture material, and surgical intervention is frequently needed in addition to medical treatment, this condition is best handled by the orthopedist.

26

Osteoporosis

Introduction..334
Senile/Postmenopausal Osteoporosis 334

Osteoporosis is a condition in which there is an absolute decrease in the amount of bone. This loss of bone mass is usually associated with aging, and is especially a problem in postmenopausal females. Because it effects trabecular bone more than compact bone, the spine, pelvis, and distal radius are especially weakened; that is why compression fractures of the vertebrae, forearm fractures and hip fractures are common sequelae. The latter, especially, are a major cause of morbidity and mortality in elderly women.

By far the most commonly seen osteoporosis is that related to aging and passing through menopause. Localized osteoporosis is also seen after prolonged immobilization, as in a cast. But keep in mind that osteoporosis can be secondary to various other conditions, which must be ruled out appropriately. These include hyperparathyroidism, Cushing's syndrome, exogenous steroid administration, myeloma, and others.

Osteoporosis must also not be confused with osteomalacia, which is inadequate mineralization of bone, or with other bone disorders such Paget's disease. The spectrum of metabolic bone disease is beyond the scope of this book; the reader is referred to an internal medicine text for furthur information.

Senile/Postmenopausal Osteoporosis
(Extremely Common)

Diagnosis:

The patient usually presents with a vertebral compression fracture or other fracture. In recent years, as awareness of the importance of preventing bone loss has spread, many asymptomatic patients are coming in asking for evaluation of their bone status and information on preventing the condition.

X-ray may suggest osteoporosis, but significant bone loss must occur before it will be apparent on plain films.

Various bone densitometry techniques are becoming more widely available, and can quantify the degree of bone loss in individual patients, thus helping therapeutic planning.

Laboratory studies including serum calcium, phosphate, alkaline phosphatese*, and SPEP are normal.

Pathophysiology

Some degree of bone loss is probably an unavoidable consequence of aging past the third or fourth decade.

But the process is often greatly accelerated in females after the menopause, especially in the several years immediately following.

The mechanism by which estrogen lack leads to bone loss is not completely elucidated as yet.

Various risk factors are, however, quite well known. These include:
- northern European or Oriental extraction
- small frame
- lack of adequate weight bearing exercise
- inadequate calcium intake
- smoking
- excessive alcohol intake

Treatment

No treatment can reverse bone loss that has already occured; but there are many factors that can help retard further loss.

In menopausal and postmenopausal women, consider the use of *estrogens*. Be aware that there are risks, contraindications and many controversies regarding their use. These will not be delved into here, as there are many issues about which no consensus has yet emerged.

Adequate *calcium* intake is very important. Aim for a total intake of one thousand milligrams daily in women on estrogen, fifteen hundred milligrams in those not on hormones. An eight-ounce glass of milk contains about three hundred milligrams; yogurt and cheese are also good sources. Supplements will often have to be recommended to achieve adequate intake. Various preparations are available; calcium carbonate is generally cheapest and most convenient, but is more likely to result in constipation. Calcium lactate (13 percent elemental calcium as compared to 40 percent for the carbonate) and calcium gluconate (9 per-

---

*Alkaline phosphatase may be increased following a fracture.

cent) are alternatives.** Kidney stones are a rare complication (24 hr. urinary calcium excretion can be checked); hypercalcemia is also rare but can occur.

Adequate doses of *Vitamin D* are necessary for calcuim absorption. The RDA of four hundred units may be assured with a miltivitamin. Higher doses have not been shown to be beneficial.

Advise the patient to get adequate *weight-bearing exercise* with daily walking.

Try to get the patient to *quit smoking* and *reduce* excess *alcohol* intake.

In severe cases other treatment including flouride, calcitonin and HCTZ may be considered; referral to an endocrinologist is best.

**Bone meal and dolomite should be avoided because they can be contaminated with lead and other toxins.

PATIENT HANDOUT 26-1

Osteoporosis

Osteoporosis is a thinning-out of bone that occurs as people get older; it is especially a problem for women after they pass the menopause.

It can lead to humped backs, broken wrists and hip fractures. But the good news is it can be slowed down and the problems it causes prevented, if you follow the following advice:

1. Be sure you get enough calcium. Try for at least one thousand milligrams a day. A glass of milk, a cup of yogurt or a one and a half ounce slice of most cheeses contain about three hundred milligrams. If you don't get enough dairy in your diet, your doctor can recommend a supplement.
2. Get enough exercise; try for a long walk every day.
3. If you smoke, quit (as if you didn't already have enough reasons to).
4. Talk to your doctor about whether it would be a good idea for you to take hormones.

27

The Temporomandibular Joint

Introduction..338
TMJ Syndrome 338

The temporomandibular joint is constructed like other joints in the body, with a lining of articular cartilage over both side and a synovial lining. Within the joint is a cartilagenous meniscus which helps to maintain proper joint congruence but can sometimes lead to difficulties.

Dysfunction of the joint and pain arising from the muscles around the joint is a widespread problem. It can be a contributing factor also for migraine headaches, muscle contraction headaches and neckaches.

The differential diagnosis of unilateral facial pain includes the following:

- temporal arteritis (which can present with "jaw claudication"; be especially careful in older patients)
- toothache
- referred pain fron the ear, sinus, parotid gland, or rarely from the eye or angina
- atypical facial pain (a neurologic condition).

TMJ Syndrome (TMJ dysfunction, myofacial pain syndrome) *(Very Common)*

Diagnosis

The patient may present with unilateral or bilateral jaw pain. The pain is often worse in the morning (when secondary to nocturnal teeth griding or jaw clenching) and is often worse after chewing or eating.

TMJ SYNDROME

Alternatively the patient may present complaining of neck pain or migraine.

It is essential to rule out other causes of unilateral facial pain as described in the introduction above.

The patient will often note that pain comes on after an increase in stress level, after dentistry, or after trauma.

The patient often complains of grinding or clicking in the TMJ, or even occasionally locking (secondary to derangement of the interarticular meniscus).

Pathophysiology

Pain is often due to a primary muscle tension state (anxiety, depression); it may secondarily lead to joint dysfunction.

Or joint dysfunction due to trauma may lead to secondary muscle tension.

Malocclusion of the teeth may be an important contributing factor.

Poor cervical posture can contribute by affecting jaw position (the lower jaw retracts with excessive cervical lordosis).

Often a vicious cycle of pain and spasm is set up. Note that with time this joint can deteriorate (i.e., degenerative arthritis) just like other joints.

Occasionally, inflammatory arthritis (like rheumatoid arthritis) may affect the TMJ.

Treatment

General measures may be all that are necessary in mild acute syndromes. These include:

- avoidance of gum chewing, clenching a pipe or cigar between the teeth, lip biting, hard foods, and wide jaw opening.
- heat (for muscle relaxation), ice (to reduce inflammation), or alternating between the two
- aspirin or other antiinflammatory medications.

If pain is primarily on arising in the morning consider the following:
- use of a cervical pillow and proper sleeping position (see Chapter 4)
- insertion of a bite block at night

If poor cervical posture is noted, advise proper posture and exercises as well as the use of a cervical pillow.

If there seems to be a large component of anxiety and/or depression, consider:

- muscle relaxants
- counseling, stress management, relaxation techniques, biofeedback
- antidepressants if indicated

If symptoms are more chronic, consider:

- referral to a physical therapist for an exercise program to restore proper jaw mechanics

- referral to a dentist for the fabrication of a customized splint to correct malocclusion
- furthur dental evaluation for adjusting occlusion by grinding or repositioning teeth.

Last resorts:

- referral for steroid injection if the joint is inflamed
- consideration of surgery after computerized tomography or an arthrogram if symptoms are severe and chronic enough to warrant, especially if symptoms like locking suggest meniscus dysfunction.

REFERENCES

GENERAL

American College of Surgeons Committee on Trauma: *Early Care of the Injured Patient.* ed 2. Philadelphia, Saunders, 1976.
Anderson JE: *Grant's Atlas of Anatomy,* ed 7. Baltimore, Williams & Wilkins, 1978.
Cailliet R: *Soft Tissue Pain and Disability.* Philadelphia, Davis, 1977.
DePalma AF: *The Management of Fractures and Dislocations: An Atlas,* 2 vols. Philadelphia, Saunders, 1970.
Cyriax J: *Testbook of Orthopedic Medicine,* 2 vols, ed 6. Baltimore, Williams & Wilkins, 1975.
Grant JBC: *Atlas of Anatomy.* ed 5. Baltimore, Williams & Wilkins, 1962.
Hollinshead WH: *Functional Anatomy of the Limbs and Back,* ed 4. Philadelphia, Saunders, 1976.
Iverson LD, Clawson DK: *Manual of Acute Orthopedic Therapeutics.* Boston, Little, Brown, 1977.
Kapandji IA: *The Physiology of the Joints,* 3 vols. Edinburgh, Churchill Livingstone, 1970.
LeVeau *et al: Biomechanics of Human Motion,* ed 2. Philadelphia, Saunders, 1977.
Ramamurti CP: *Orthopedics in Primary Care.* Baltimore, Williams & Wilkins, 1979.
Rockwood CA, Green D P: *Fractures,* 2 vols. Philadelphia, Lippincott, 1975.
Turek SL: *Orthopedics: Principles and their Application,* ed 3. Philadelphia, Lippincott, 1977.

CHAPTER 2

Daniels L, Worthingham C: *Muscle Testing: Techniques of Manual Examination,* ed 4. Philadelphia, Saunders, 1980.
Beetham WP *et al: Physical Examination of the Joints.* Philadelphia, Saunders, 1965.

Edeiken J, Hodes PJ: *Roentgen Diagnosis of Diseases of the Bone,* 2 vols, ed 2. Baltimore, Williams & Wilkins, 1973.
Forrester DM, Brown JC, Nesson JW: *The Radiology of Joint Disease,* Philadelphia, Saunders, 1978.
Licht S (ed): *Electrodiagnosis and Electromyography,* ed 3. New Haven, Elizabeth Licht, 1971.
Murray RO, Jacobsen HG: *The Radiology of Skeletal Disorders,* 4 vols, ed 2. Edinburgh, Churchill Livingstone, 1977.
Watanabe M, Takeda S, Ikeuchi H: *Atlas of Arthroscopy,* ed 3. Tokoyo, Igaku-Shoin, 1979.

CHAPTER 3

Baker CE: *Physicians Desk Reference.* ed 39. Oradell, New Jersey, Medical Economics Co, 1985.
Baker CE: *Physicians Desk Reference for Nonprescription Drugs,* ed 1. Oradell New Jersey, Medical Economics Co, 1980.
Rusk HA: *Rehabilitation Medicine,* ed 4. St. Louis, Mosby, 1977.
Steinbrocker O, Neustadt DH: *Aspiration and Injection Therapy in Arthritis and Musculoskeletal Disorders,* Hagerstown, Maryland, Harper, 1972.

CHAPTER 4

Cailliet R: *Neck and Arm Pain.* Philadelphia, Davis, 1964.
Bateman JE: *The Shoulder and Neck,* ed 2. Philadelphia, Saunders, 1978.
Jackson R: *The Cervical Syndrome,* ed 3. Springfield, Illinois, Charles C Thomas, 1971.

CHAPTER 5

Calliet R: *Neck and Arm Pain.* Philadelphia, Davis, 1964.
Chusid JG: *Correlative Neuroanatomy and Functional Neurology,* ed 15. Los Altos, California, Lange, 1973.
Hale MS: *A Practical Approach to Arm Pain,* Springfield, Illinois, Charles C Thomas, 1971.
Kopell HP, Thompson WAL: *Peripheral Entrapment Neuropathies.* Huntington, Robert E. Krieger, New York, 1976.

CHAPTER 6

Bateman JE: *The Shoulder and Neck,* ed 2. Philadelphia, Saunders, 1978.
Cailliet R: *Shoulder Pain.* Philadelphia, Davis, 1966.

CHAPTER 7

Lister GD, Belsole RB, Keinert HE: The radial tunnel syndrome. *J Hand Surg* 4(1):52, 1979.

REFERENCES

CHAPTER 8

Cailliet R: *Hand Pain and Impairment*. Philadelphia, Davis, 1975.

CHAPTER 9

Cailliet R: *Hand Pain and Impairment*. Philadelphia, Davis, 1975.
Ramamurti CP: *Orthopedics in Primary Care*. Baltimore, Williams & Wilkins, 1979.
Ruby LK: *Common Hand Injuries in the Athlete,* Symposium on Sports Injuries, Orthopedic Clinics of North America, 11, 4. Philadelphia, Saunders, 1980.

CHAPTER 11

Cailliet R: *Low Back Pain Syndrome,* ed 2. Philadelphia, Davis, 1968.
DePalma Af, Rothman RH: *The Intervertebral Disc*. Philadelphia, Saunders, 1970.
Drugs for postmenopausal osteoporosis. *Medical Letter on Drugs and Therapeutics* 22(11), 1980.
Helfet AJ, Gruebel Lee DM: *Disorders of the Lumber Spine*. Philadelphia, Lippincott, 1978.
MacNab I: *Backache*. Baltimore, Williams & Wilkins, 1977.
McKenzie RA: *The Lumbar Spine,* Waikanae NZ, Spinal Publications, 1981.

CHAPTER 13

Chusid JG: *Correlative Neuroanatomy and Functional Neurology,* ed 15. Los Altos, California, Lange, 1973.
Cailliet R: *Foot and Ankle Pain*. Philadelphia, Davis, 1968.
Kopell HP, Thompson WAL: *Peripheral Entrapment Neuropathies*. Huntington, New York, Robert E. Krieger, 1976.

CHAPTER 14

Strange FGSC: *The Hip*. Baltimore, Williams & Wilkins, 1965.

CHAPTER 15

Cailliet R: *Knee Pain and Disability*. Philadelphia, Davis, 1973.
Helfet AJ: *Disorders of the Knee*. Philadelphia, Lippincott, 1974.
Smillie IS: *Injuries of the Knee Joint,* ed 5. Edinburgh, Churchill Livingstone, 1978.

CHAPTER 17

Cailliet R: *Foot and Ankle Pain*. Philadelphia, Davis, 1968.

CHAPTER 18

Cailliet R: *Foot and Ankle Pain*. Philadelphia, Davis, 1968.
Ramamurti CP: *Orthopedics in Primary Care*. Baltimore, Williams & Wilkins, 1979.

CHAPTER 19

American Academy of Pediatrics: *School Health: A Guide for Health Professionals*. 1977.
Green M, Haggerty RJ: *Ambulatory Pediatrics II*. Philadelphia, Saunders, 1977.
Hoekelman R et al: *Principles of Pediatrics: Health Care of the Young*. New York, McGraw-Hill, 1978.
Lovell WW, Winter RB (eds): *Pediatric Orthopedics,* 2 vols. Philadelphia, Lippincott, 1978.
Symposium of common orthopedic problems, *Pediatric Clinics of North America*. Philadelphia, Saunders, 1977.
Vaughn VC, McKay RJ: *Nelson's Textbook of Pediatrics*. ed 10. Philadelphia, Saunders, 1975.
Shaw R: *Pitching,* Chicago, Contemporary Books, 1972.

CHAPTER 20

Anderson B: *Stretching*. Bolinas, California, Shelter Publications, 1980.
Brody DM: *Running Injuries*. Ciba Clinical Symposia, Vol 32(4), 1980.
Clancy WG (ed): Symposium on runner's injuries, *Am J Sportsmed* 8:2, 4, 1980.
Daniel DM: *Lower Limb Maladies in the Runner*. A Course on Orthopedics for the Primary Physician, UCSD, San Diego, March 1980.
James SL, Bates BT, Osternig LR: Injuries to runners, *Am J Sportsmed* 8:5, 1980.
Mann RA, Hagy J: Biomechanics of walking, running and sprinting, *Am J Sportsmed* 8(5), 1980.
Pagliano J, Jackson D: The ultimate study of running injuries, *Runner's World* 15:11, 1980.
Kulund D: *The Injured Athlete,* Philadelphia, JB Lippincott, 1982.
Roy S, Irvin R: *Sports Medicine,* Englewood Cliffs, Prentice Hall, 1983.

CHAPTER 21

Adams RD: *Diseases of Muscle,* ed 3. Hagerstown, Maryland, Harper, 1975.

CHAPTER 22

Arthritis Foundation. Primer on the Rheumatic Diseases, ed 7. New York, 1973.
Katz WA: *Rheumatic Diseases: Diagnosis and Management*. Philadelphia, Lippincott, 1977.
Kelly WN et al: *Textbook of Rheumatology*. Philadelphia, Saunders, 1981.
McCarty DJ: *Arthritis and Allied Conditions: A Textbook of Rheumatology,* ed 9. Philadelphia, Lea & Febiger, 1979.
Wintrobe MM et al: *Harrisons Principles of Internal Medicine,* ed 7. New York, McGraw-Hill, 1974.

REFERENCES

CHAPTER 23

Arthritis Foundation, Primer on the Rheumatic Diseases, ed 7. New York, 1973.
Katz WA: *Rheumatic Diseases: Diagnosis and Management*. Philadelphia, Lippincott, 1977.
Kelly WN et al: *Textbook of Rheumatology*. Philadelphia, Saunders, 1981.
McCarty DJ: *Arthritis and Allied Conditions: A Textbook of Rheumatology,* ed 9. Philadelphia, Lea & Febiger, 1979.
Wintrobe MM et al: *Harrisons Principles of Internal Medicine*, ed 7. New York, McGraw-Hill, 1974.

CHAPTER 24

Arthritis Foundation. Primer on the Rheumatic Diseases, ed 7. New York, 1973.
Baker CE: *Physicians Desk Reference*, ed 39. Oradell, New Jersey, Medical Economics Co, 1985.
Katz WA: *Rheumatic Diseases: Diagnosis and Management*. Philadelphia, Lippincott, 1977.
Kelly WN et al: *Textbook of Rheumatology*. Philadelphia, Saunders, 1981.
McCarty DJ: *Arthritis and Allied Conditions: A Textbook of Rheumatology* ed 9. Philadelphia, Lea and Febiger, 1979.
Talbott JH, Yu T-F: *Gout and Uric Acid Metabolism*. New York, Stratton, 1976.
Wintrobe MM et al: *Harrisons Principles of Internal Medicine*, ed 7. New York, McGraw-Hill, 1974.

CHAPTER 25

Gardner P, Provine HT: *Manual of Acute Bacterial Infections*. Boston, Little, Brown, 1975.
Iverson Ld, Clawson DK: *Manual of Acute Orthopedic Therapeutics*. Boston, Little, Brown, 1977.

Index

Note: f = figure; n = footnote; t = table

Abdominal strengthening, 145
Abductor exercises, 203
Accessory bones, 250
Accommodative padding, 237f
Acetaminophen, 26
Achilles tendon
  pain in area, 220–223, 293
  strain
    diagnosis, 223
    pathophysiology, 223
    treatment, 223
  stretching, 303
  tendonitis
    diagnosis, 221, 293
    pathophysiology, 223
    site of tenderness, 211f, 222f
    treatment, 223, 298t
Acid phosphatase, 15
Acromioclavicular joint
  arthritis
    diagnosis, 60, 67
    pathophysiology, 68
    sites of tenderness, 61f
    treatment, 68
  pain, 65
  sprain
    diagnosis, 71
    pathophysiology, 71
    treatment, 71

Adduction, of foot in infant, 259f
Adductor
  strain, diagnosis, 295
  stretching, 302
Adhesive capsulitis
  diagnosis, 60, 65, 68
  pathophysiology, 68
  prevention of, 65, 66
  site of tenderness, 61f
  treatment, 68
Adson's maneuver, 50f, 51
Aerobic exercise, 17
Alcaptonuria, 324
Alkaline phosphatase, 15
Allopurinol, 327t
Amphiarthrodal joint, 3
Analgesics
  nonsteroidal antiinflammatory medications as, 23t–24t
  pain relief and, 26
Anesthetizing, of finger, 107f
Ankle
  anatomy, 209f, 210f
  degenerative arthritis
    diagnosis, 215
    site of tenderness, 21f
    treatment, 216
  instability and locking
    diagnosis, 220

347

evaluation, 219–220
   treatment, 220
 lumps, 220
 pain
   differential diagnosis, 293
   examination, 211–212
   history, 210
   sites of tenderness, 211f
 pathophysiology, 220
 patient handouts
   how to tape your ankle, 228
   if you keep spraining your ankle, 226–227
   your sprained ankle, 224–225
 sprain
   diagnosis, 218
   examination, 216–217
   pathophysiology, 218
   patient handouts, 224–227
   treatment, 218–219
 swelling around, 213f
 tenderness, 211f
 treatment of runner's injuries, 297t–298t
Ankylosing spondylitis
 diagnosis, 137–138
 history of, 122
 treatment, 138
Annulus fibrosis
 anatomy, 115f
 degenerative changes, 116f, 128
Anserine tendonitis
 diagnosis, 190
 injection of insertion, 190f
 pathophysiology, 190
 site of tenderness, 175f
 treatment, 190f
Anterior scalene syndrome, diagnosis, 51
Anterior tibialis tendonitis
 diagnosis, 214, 293
 pathophysiology, 214
 site of tenderness, 211f
 treatment, 214, 297t
Antidepressants, 27–28
Antiinflammatory medications
 choosing, 21–22t
 contraindications and precautions, 20–21
 marketed as analgesics, 23t–24t
 nonsteroidal, 23–24t, 26
Antinuclear antibodies, 15
Apophysis, definition, 2
Apophysitis, definition, 3
Arthritis. *See also* specific type of arthritis

definition, 3
 types affecting joints of hand and fingers, 98
Arthrography, 11–12
Arthroscopy, 12
Articular cartilage, 3
Articular pain, vs. periarticular and referred pain, 5–6
Artistospan, 25t
Aspiration technique
 for ankle joint, 212f
 for ganglion, 93f
 for knee joint, 173f
 for shoulder joint, 69f
 for subtalar joint, 215f
 for wrist joint, 92f
Aspirin
 disadvantages, 21
 for pain relief, 26
 types of preparations, 22t
Avascular necrosis, definition, 3
Avulsion fracture(s)
 definition, 3
 of hand, 100f

Back. *See also* Low back; Upper back
 abnormal curvature, 273–275f
 acute strains, diagnosis, 272
 dislocation, 140
 fractures, 140
 pain in children and adolescents, 272–273
Backknee, evaluation and management, 267–268f
Bacterial endocarditis, 324
Baker's cyst
 diagnosis, 193, 194, 206, 295
 knee swelling, 172f
 pathophysiology, 194–195
 treatment, 195
Bankart lesion, 70
Baseball finger
 diagnosis, 106
 management, 107f
Battered child syndrome, 258, 277
Behcet's syndrome, 324
Benign neoplasms, low back pain and, 139
Benzodiazepines, 26
Biceps tendonitis
 diagnosis, 65, 66
 injection, 67f
 pathophysiology, 66
 sites of tenderness, 61f
 treatment, 66–67

INDEX

Biofeedback training, 28
Bite/tooth wound, human, 108
Bladder symptoms, lumbar derangement with, 133
Blisters
 diagnosis, 289
 pathophysiology, 290
 treatment, 290
Blood tests, 14–15
Bone, definition, 2–3
Bone infection. *See* Infection, bone
Bone scanning
 in low back pain evaluation, 122
 of stress fracture, 285
 technique, 12–13
Bouchard's nodes, 98, 315
Boutonniere deformity, 103, 105
Bowel symptoms, lumbar derangement with, 133
Bowleg, evaluation and management, 267–268*f*
Brachial plexitis
 diagnosis, 50
 pathophysiology, 53
 treatment, 53
Brain lesion, diagnosis, 45, 150
Breastrokers Knee
 diagnosis, 187
 pathophysiology, 187
 treatment, 188
Bruise. *See* Contusion; specific bruise
Bunion
 diagnosis, 240
 site of tenderness, 240*f*
Bunionette, 240
Bursa(e)
 definition, 3
 infected superficial
  diagnosis, 331
  pathophysiology, 331
  treatment, 332
Bursitis, definition, 4. *See also* specific bursitis
Buttock pain, 131, 295

C6 root syndrome, diagnosis, 47*f*
C7 root syndrome, diagnosis, 47*f*
C8 root syndrome, diagnosis, 47*f*
Calcaneal bursitis, posterior
 diagnosis, 221
 pathophysiology, 221
 treatment, 221

Calcaneo valgus, diagnosis, 259*f*
Calcific tendonitis
 diagnosis, 64–65
 pathophysiology, 65
 treatment, 66
Calcium
 osteoporosis and, 335–336
 serum levels, 15
Calf
 exercises, 203
 measuring tightness, 182, 184*f*
 pain in, 206, 293
 pull, diagnosis and treatment, 207
 strain
  diagnosis, 293, 295
  treatment, 298*t*
 stretching, 303
Callus
 diagnosis, 248, 249*f*
 pathophysiology, 248
 patient self-care, 255
 treatment, 248
Capsule, of joint, 3
Capsulitis. *See* Adhesive capsulitis
Carisoprodol (Rela; Soma), 27*t*
Carpal tunnel syndrome
 diagnosis, 45, 49, 51, 53–54
 pathophysiology, 54
 treatment, 54
Carpal-metacarpal joint, degenerative arthritis, 88*f*
Cartilage
 articular, 3
 of knee, torn. *See* Meniscus tears
Cauda equina compression, 133
CBC, 14
Celestone soluspan, 25*t*
Cellulitis
 diagnosis, 332
 pathophysiology, 332–333
 treatment, 333
Central extensor slip tear
 diagnosis, 102, 103, 104
 management, 105
Cervical collar, patient use, 40
Cervical radiculitis
 diagnosis, 32, 48, 51
 pathophysiology, 52
 treatment, 52
Cervical spine anatomy, 32*f*. *See also* Neck; Upper back
Cervical sprain

diagnosis, 32, 37–38
treatment, 38
Charcot joints, 315
Chest pain, musculoskeletal
  diagnosis, 109–110
  evaluation, 109
  pathophysiology, 110
  treatment, 110
Chin tucks, 43
Chlorzoxazone (Paraflex), 27f
Chondrocalcinosis, 328
Chondromalacia, see Patellofemoral pain, 174f
Chordoma, 139
Chronic ligamentous instability
  diagnosis, 191, 192
  pathophysiology, 192
  treatment, 192
Clubfoot, diagnosis, 258
CNS lesion, diagnosis, 45, 48
Coccydynia
  diagnosis, 149
  pathophysiology, 149
  treatment, 149
Coccyx
  bruise or fracture
    diagnosis, 148–149
    treatment, 149
  coccydynia, 149
  evaluation, 148
Cock-up wrist splint, 54
Colchicine, 24t
Cold application
  technique, 19
  use by patient, 29–30
Collateral ligament(s)
  injury, diagnosis, 195
  sprain
    diagnosis, 103, 105, 195–196
    treatment, 105, 196
  stability, 176, 178f
Common peroneal nerve compression syndrome
  diagnosis, 152f, 157
  pathophysiology, 157
  treatment, 157
Compartment syndrome(s)
  anterior recurrent, diagnosis, 293
  definition, 4, 312–313
  diagnosis, 151, 154f
  pathophysiology, 313
  treatment, 313
Complete blood count (CBC), 14

Compression fracture of vertebrae
  diagnosis, 117, 136
  treatment, 136
Computerized tomography, 12
Congenital hip dysplasia, evaluation, 260–262f
Congenital muscular torticollis, diagnosis, 275
Congenital subluxability of hip
  diagnosis, 260, 262
  pathophysiology, 262
  treatment, 262
Connective tissue disease, definition, 4
Contractures, hand, 106
Contusion
  definition, 4
  diagnosis, 308
  pathophysiology, 309
  treatment, 310
Coracoacromial arch, 65
Corn(s)
  soft
    diagnosis, 248, 249f
    pathophysiology, 248
    treatment, 248
  hard
    diagnosis, 248, 249
    pathophysiology, 248
    treatment, 250
  patient self-care, 255
Costochondritis
  diagnosis, 109–110
  treatment, 110
Costoclavicular maneuver, 50f, 51
Costoclavicular syndrome
  diagnosis, 51, 52
  pathophysiology, 52
  treatment, 52–53
Cramp, definition, 4
Cruciate ligament(s)
  anterior, injury
    diagnosis, 195, 196
    pathophysiology, 196–197
    treatment, 197
  integrity testing, 176, 179f
  posterior, injury
    diagnosis, 197
    treatment, 197
Crystalline disease(s)
  definition, 4
  gout, 325–326
  monoarthritis and, 6
  pseudogout, 326, 328

INDEX

CT scanning, 12
Cubital tunnel syndrome, 55
Cyclobenzaprine (Flexeril), 27t

Decadron, 25t
Deep peroneal nerve compression syndrome
  diagnosis, 152f, 157
  pathophysiology, 157
  treatment, 157
Degenerative arthritis
  of ankle
    diagnosis, 215
    site of tenderness, 211f
    treatment, 216
  of carpometacarpal joint of thumb
    diagnosis, 90
    pathophysiology, 90
    treatment, 91
  clinical features, 314–315
  definition, 4
  diagnosis, 191, 315–316
  diffuse, of foot, 235
  of elbow
    diagnosis, 79–80
    sites of tenderness, 77f
  of first carpal-metacarpal joint, 88f
  of first MTP joint, 240f
    diagnosis, 241, 291
    pathophysiology, 241
    treatment, 241–241f, 297f
  of glenohumeral joint
    diagnosis, 65, 69–70
    pathophysiology, 70
    treatment, 70
  of hip
    diagnosis, 166
    treatment, 166
  of knee
    diagnosis, 176, 182, 190–191
    pathophysiology, 191
    site of tenderness, 174f
    treatment, 191
  of lower back, 134
  monoarthritis and, 6
  pathophysiology, 316
  patient handout, 318–319
  of shoulder, diagnosis, 60
  of subtalar joint
    diagnosis, 214–215
    pathophysiology, 215
    treatment, 215
  treatment, 316–317
  of wrist
    diagnosis, 91
    pathophysiology, 91
    treatment, 92
Degenerative disc disease, 115f
Degenerative joint disease. *See* Degenerative arthritis
Depo-medrol, 25t
DeQuervain's tendonitis
  diagnosis, 88–89
  injection for, 90f
  pathophysiology, 89
  site of tenderness, 88f
  treatment, 89
Diabetes, 45
Diabetic mononeuropathy, 7
Diagnosis. *See also* specific conditions
  arthrography, 11–12
  blood tests, 14–15
  bone scanning, 12–13
  CT scanning, 12
  EMG and NCB, 13
  history, 9–10
  myelography, 12
  physical examination, 10
  synovial fluid analysis, 13–14t
  x-rays, 11
Diarthrodal joint, 3
Diathermy, 20
Diflunisal, 23t
DIP joint sprains, evaluation, 105–107f
Disc, intervertebral disc
  degenerative disease, 115f
  derangements, diagnosis, 117
  as model for low back pain, 127–128
  rupture or herniation, 128
  stress and deterioration, 126
Discitis
  diagnosis, 138, 272
  pathophysiology, 139
  treatment, 139
Dislocation, definition, 4. *See also* specific dislocation
Distal interphalangeal joint sprains, evaluation, 105–106
Distal neurologic status, 10
Donut pad, 237f
Drainage, of knee joint, 194
Drawer tests, 176, 179f, 217f

Dropped finger
    diagnosis, 106
    management, 107f
Drug-induced lupus-like syndromes, 15
Duck walk maneuver, 176, 183f
Dupuytren's contracture, 107
Dynamogram, 134
Dysfunction. *See* Lumbar dysfunction

Eating, precautions for low back pain, 144
Effusion
    ankle, 213f
    elbow, 78f
    knee, 172f, 172–173
Ehlers-Danlos syndrome, 324
Elbow
    anatomy, 75f
    aspiration or injection technique, 80f
    degenerative arthritis
        diagnosis, 79–80
        sites of tenderness, 77f
    pain
        degenerative arthritis, 79–80
        examination, 76
        history, 74, 76
    problems in children
        Little League elbow, 276–277
        radial head subluxation, 276
    sites of tenderness, 77f
    sprains
        diagnosis, 81
        treatment, 82
    swelling
        differential diagnosis, 78f
        evaluation, 80
    trauma, 81–82
    x-ray, fat pad sign on, 82f
Electrodiagnosis, 51, 154
Electromyography (EMG), 13
Electrophoresis, serum protein, 15
Electrotherapy, 20
EMG (electromyography), 13
Enteropathic arthritis
    clinical features, 323
    HLA-B27 antigen and, 15
    treatment, 323
Epicondylitis. *See* Lateral epicondylitis; Medial epicondylitis
Epiphyseal injury
    definition, 4
    Salter's classification, 258t

Epiphyseal plate, definition, 3
Equinovarsus, 259f
Erosive osteoarthritis, 315
Erythrocyte sedimentation rate (ESR), 124, 138
Exercise(s)
    aerobic, 17
    flexibility, 17–18
    knee, 201–203
    low back, 145
    for low back pain, 144
    neck, for patient, 43
            for rehabilitation in epicondylitis, 83
    shoulder mobilization, 72–73
    strengthening, 18, 145
    stretching, 301–303
    Williams', 127
Extension, 145
Extensor tendonitis
    diagnosis, 214, 246, 292
    pathophysiology, 214, 246
    site of tenderness, 88, 211f
    treatment, 214, 246, 297t

Facet joint
    anatomy, 115f
    syndrome, diagnosis, 117, 134
Familial Mediterranean fever, 324
Felon, 332
Felty's syndrome, 321
Femoral anteversion, excessive
    diagnosis, 266f
    treatment, 267
Femoral head, avascular necrosis. *See* Legg-Perthe's disease
Femoral stress fracture, diagnosis, 295
Femoral version, evaluation of, 266f
Fenoprofen, 23t
Fibromyalgia. *See* Fibrositis
Fibrositis
    diagnosis, 33, 310
    pathophysiology, 310
    treatment, 311
Finger
    anatomy, 96f, 97f
    anesthezing, 107f
    mallet or dropped
        diagnosis, 106
        management, 107f
    sprains, of MCP joint, 102
Finkelstein's test, 88f, 89

INDEX
353

Flat feet. *See* Foot strain in the hyperpronating foot, chronic
Flexeril (cyclobenzaprine), 27*t*
Flexibility exercises, 17–18
Flexion, 145
Foot. *See also* Forefoot; Heel pain
 anatomy, 231*f*
 deformities in infants, 259*f*
 diffuse pain
  acute foot strain, 233
  chronic foot strain in the hyperpronating foot, 233–234
  evaluation, 231–232
  examination, 230
 flexible pronated (flat) foot in infant and children
  diagnosis, 263
  pathophysiology, 263
  treatment, 263
 history, 230
 keratoses, 248–250*f*
 lumps, 250
 pain in and around the ball
  differential diagnosis in runners, 291
  evaluation, 239
  hallux valgus and bunion, 239–241*f*
  treatment of runners injuries, 297*t*
 pain in bottom of heel and arch, 292–293*f*
  treatment of runners injuries, 297*t*
 pain in children and adolescents
  evaluation, 264
  Freiberg's disease, 264
  Kohler's disease, 264–265
 pain on top (dorsal)
  differential diagnosis, 292
  extensor tendonitis, 246
  stress fracture of metatarsal bone, 245–246
  treatment of runners injuries, 297*t*
 patient handouts
  taking care of calluses and corns, 255
  what to look for in a shoe, 254
 rigid pronated in infant and children, diagnosis, 263
 sites of tenderness of ball, 240*f*
 sprain
  diagnosis, 251
  treatment, 251
 strain, acute
  diagnosis, 233
  treatment, 233

strain, chronic
 due to other deformities, 235
 in hyperpronating foot, 233–234
 in rigid, high-arched foot (pes cavus), 235
stretching your infant's, 280
therapeutic devices, 237*f*
transverse arch pain
 differential diagnosis, 292
 evaluation, 242, 243*f*
 metatarsalgia, 244
 Morton's neuroma, 244–245
 treatment of runners injuries, 297*t*
trauma, 250–253*f*
Foramenal compression, 50*f*, 51
Forefoot
 adduction in infant, 259*f*
 sites of tenderness, 243*f*
Fracture(s). *See also* specific fractures
 avulsion, definition, 3
 definition, 4
 pathologic, definition, 4–5
 types in hand and fingers, 99–100
Freiberg's disease
 diagnosis, 264
 pathophysiology, 264
 treatment, 264
Frozen shoulder. *See* Adhesive capsulitis
Functional status, 10

Gait, intoeing
 evaluation, 265–266*f*
Gamekeeper's thumb, 103, 104*f*
Ganglia
 ankle, 220
 diagnosis, 92–93, 108
 foot and ankle, 250
 pathophysiology, 93
 treatment, 93*f*
Gaucher's disease, 324
Generalized myositis ossificans, 309*n*
Genu recurvatum, 267, 268*f*
Genu valgum, 267, 268*f*
Genu varum, 267, 268*f*
Gerdy's tubercle, site of tenderness, 175*f*
Glenohumeral joint, degenerative arthritis
 diagnosis, 65, 69–70
 pathophysiology, 70
 treatment, 70
Gonococcal arthritis, 6
 diagnosis, 330
 treatment, 331

354 INDEX

Gout
  clinical features and diagnosis, 325–326
  diagnosis, 239
  pathophysiology, 326
  treatment, 326, 327t
Gouty arthritis, chronic, 325
Groin
  pain, 295
  pull
    diagnosis, 165
    pathophysiology, 165
    site of tenderness, 161f
    treatment, 165
    strain, 299t
    stretching, 302
Ground reaction force
  measures to reduce, 296t
  overuse injuries and, 287

Hallux valgus
  diagnosis, 239–240
  pathophysiology, 241
  site of tenderness, 240f
  treatment, 241
Hammertoes
  diagnosis, 234f, 246
  pathophysiology, 246
  treatment, 246
Hamstring(s)
  exercises, 202
  measuring tightness, 182, 184f
  pull
    diagnosis, 166
    pathophysiology, 166
    site of tenderness, 161f
    treatment, 166
  strain, diagnosis, 295
  stretching, 302
Hand
  anatomy, 96f, 97f
  contractures, 106
  dislocations, 101–102
  fractures
    avulsion, 100t
    tuft of distal phalanx, 100–101
    types, 99–100
  infections, 107f, 332
  lumps, evaluation, 107–108
  pain
    atraumatic pain in joints and fingers, 98–99
    diffuse nonlocalized, 98
Heat application
  technique, 18–19
  use by patient, 29
Heberden's nodes, 98, 315
Heel pain
  of cord, in children, 265
  evaluation, 236
  pad contusion
    diagnosis, 292
    treatment, 297t
  pad, painful (stone bruise)
    diagnosis, 238
    pathophysiology, 239
    site of tenderness, 238f
    treatment, 239
  plantar fasciitis, 236–238
  Sever's disease of, 3, 265
  sites of tenderness, 238f
Heel spur, 236–237
Heel wedge, 237f
Hemarthrosis, 257
Herpes zoster, preeruptive, 7
Hexadrol, 25t
Hill-Sachs lesion, 70
Hip
  anatomy, 160f
  congenital dysplasia, evaluation, 260–262f
  congenital subluxability, 262
  degenerative arthritis
    diagnosis, 166
    treatment, 166
  evaluation, 159–160
  examination, 161, 163
  Legg-Perthe's disease of, 3
  pain and strain
    history, 160–161
    trochanteric bursitis, 163–164
  pathology, Patrick's test to rule out, 121f
  pointer
    diagnosis, 165
    pathophysiology, 165
    site of tenderness, 161f
  treatment, 165
  range of motion, 162f
  sites of tenderness, 161f
  transient synovitis, 270
  x-ray, 163
HLA-B27 antigen, 15, 125, 138
Housemaid's knee. *See* Prepatellar bursitis

INDEX 355

Hyperabduction maneuver, 51
Hyperacute shoulder, 64
Hyperlordosis, 119f
Hyperpronation, measures to control, 296t
Hyperventilation, diagnosis, 48
Hypouricemic drugs, 327t

Ibuprofen, 23t
Iliac pain, 295
Iliotibial band tendonitis, site of tenderness, 175f
Immobilization, 17
Impingement syndrome (rotator cuff tendonitis)
  diagnosis, 64–65
  injection, 67f
  pathophysiology, 65
  site of tenderness, 61f
  treatment, 66
Indomethicin, 23t
Infection
  bone, 333
  hand, 108f
  joint
    gonococcal arthritis, 330–331
    septic arthritis, 329–330
    tuberculous, mycotic and syphilitic arthritis, 331
    viral arthritis, 331
  soft tissue
    cellulitis, 332–333
    of hand, 332
    infected superficial bursae, 331–332
    lymphangitis, 333
    septic tenosynovitis, 332
Inflammatory process, 10
Informed consent, 24
Infrapatellar bursa swelling, 172f, 193
Infrapatellar bursitis
  diagnosis, 194
  pathophysiology, 194
  treatment, 194
Infrared, 19
Ingrown toenail
  diagnosis, 247
  pathophysiology, 247
  treatment, 247–248
Injection technique
  for ankle joint, 212f
  carpal tunnel, 54f
  of CMC joint of thumb, 91f
  for cuff tendonitis and subacromial bursitis, 67f
  for DeQuervain's tendonitis, 90f
  for knee joint, 173f
  for subtalar joint, 215f
  trigger finger, 100f
  of trigger points, 36f
  wrist joint, 92f
Interdigital neuroma. See Morton's neuroma
Interosseous ligament disruption, 218f
Intervertebral disc. See Disc, intervertebral
IP joint of small toe, dislocation, 252–253f
Ischeal bursitis
  diagnosis, 164
  pathophysiology, 164
  treatment, 164
Isometrics, crossed leg, 201
ITB tendonitis, diagnosis, 295

Joint. See also specific joint
  definition, 3
  degenerative disease, definition, 4
  inflammation. See Arthritis
Joint effusion. See Effusion
Joint infection. See Infection, joint
"Joint mouse", 192
Jones' criteria for diagnosis of Rheumatic fever, 279
Jones fracture. See Metatarsal, fracture of base of fifth
Juvenile rheumatoid arthritis
  clinical features, 278
  pathophysiology, 278
  treatment, 278

Kenalog, 25t
Keratoconjunctivitis sicca, 321
Keratoderma blennorrhagica, 324
Keratoses of foot, 248–250f
Knee. See also Patella(r)
  anatomy, 169f–171f
  aspiration or injection technique, 173f
  breaststrokers, 53
  degenerative arthritis
    diagnosis, 176, 182, 190–191
    pathophysiology, 191
    site of tenderness, 174f
    treatment, 191
  examination, 171–182f, 186t
  exercises, 201–203

giving way and catching/locking, 191–193
history, 168
Osgood-Schlatter's disease of, 3
pain
   in children and adolescents, 268–269
   differential diagnosis in runners, 293, 295
   evaluation, 182
   sites of tenderness, 174f
   treatment of runner's injuries, 298t–299t
sprain. *See* Collateral ligament, sprain
swelling, 172f
   evaluation, 193
trauma, evaluation, 195
Knockknee, evaluation and management, 267–268f
Kohler's disease
   diagnosis, 264
   pathophysiology, 264
   site of tenderness, 211f
   treatment, 265
Kyphosis
   diagnosis, 199f
   with lumbar derangement, 130–131

L3 root syndrome, 152f
L4 root syndrome, 153f
L5 root syndrome, 153f
Lachman's test, 176, 180f
Larsen-Johannson's disease, 269
Lateral epicondylitis (tennis elbow)
   diagnosis, 76
   pathophysiology, 76
   patient handout, 84–85
   rehabilitation exercises, 83
   site of tenderness, 77f
   treatment, 76, 78–79
Lateral femoral cutaneous nerve syndrome, 152f
Lateral ligaments of ankle, 216f
LE cell preparation, 15
Leg. *See also* Calf; Lower extremity
   measuring length, 120f
   pain down. *See also* Sciatica
      lumbar derangements with, 131–132
Legg-Perthe's disease, 3
   diagnosis, 271
   pathophysiology, 271
   treatment, 271
Lifting and carrying, 144
Ligament
   definition, 3

Ligamentous instability, chronic of knee
   diagnosis, 191, 192
   pathophysiology, 192
   treatment, 192
Little league elbow
   diagnosis and pathophysiology, 276
   management, 277
Low back
   anatomy, 115f, 116f
   reminder, 114
   pain
      causes, 126–127
      chronic, diagnosis of, 133
      degenerative arthritis and, 134
      evaluation, 114, 117
      examination, 118–120
      exercises, patient handout, 145
      fractures, dislocations and, 140
      history, 117–118
      muscle strain diagnosis and, 117
      other lab tests, 124–125
      posture and precautions, 143–144
      self care patient handout, 141–142
      syndrome, 129
      treatments, 133, 146t–147t
      unified model, 127–128
   strain, 126, 135
Lower extremity. *See also* specific syndromes
   neurologic symptoms and diffuse pain
      evaluation, 150–155
      syndromes, 155–158
      strength testing, 155t
Lumbar derangement, 130
   with acute severe pain, 132
   with bowel or bladder symptoms, 133
   with kyphotic deformity, 130–131
   with mild weakness, 132
   with pain radiated to buttock, 131
   pathophysiology, 130
   prolonged, 133
   recurrent, 133
   with scoliotic deformity, 131
   with significant or worsening weakness, 132
   treatment, 130
Lumbar dysfunction
   diagnosis, 129
   pathophysiology, 129
   treatment, 129
Lumbar lordosis, 115f
   loss of, 119f
   low back pain and, 126–127

INDEX

Lumbar sprain
  diagnosis, 117, 135
  pathophysiology, 135
  treatment, 135
Lumbar vertebra, 116*f*
Lumbosacral spine. *See also* Lower back
  anatomy, 115*f*, 116*f*
Lymphangitis
  diagnosis, 333
  treatment, 333

Malignancy. *See also* Neoplasm; specific malignancy
  causing low back pain, 117
Mallet finger
  diagnosis, 106
  management, 107*f*
Management of musculoskeletal disorders, 1, 7–8
Maneuvers, in defining upper extremity problem, 50*f*
Manipulation, 20
March fracture. *See* Stress fracture of metatarsal bone
Massage, 20
McMurray's test, 176, 181*f*, 182*f*
MCP joint sprain, of thumb, 102–103
Meclofenamate, 23*t*
Medial distial tibia, tenderness along, 293
Medial epicondylitis
  diagnosis, path and treatment, 79
  rehabilitation exercises, 83
  site of tenderness, 77*f*
Median nerve compression syndrome
  diagnosis, 46*f*
  in forearm. *See* Pronator teres syndrome
  in wrist. *See* Carpal tunnel syndrome
Meniscus
  painful, 174*f*
  tears
    diagnosis, 182, 188, 192, 195
    pathophysiology, 188
    treatment, 189
Meralgia paresthetica
  diagnosis, 156
  pathophysiology, 156
  treatment, 157
Metacarpal joint sprain, of thumb, 102–103
Metastatic tumors, low back pain and, 139–140
Metatarsal
  fracture in base of fifth, 251–252

  stress fracture
    diagnosis, 292
    treatment, 246, 297*t*
Metatarsal phalangial joint
  degenerative arthritis, 240*f*
  diagnosis, 241, 291
  pathophysiology, 241
  treatment, 241–242*f*, 297*t*
  injection into first, 242*f*
Metatarsalgia
  diagnosis, 244, 292, 292*f*
  pathophysiology, 244
  treatment, 244, 297*t*
Metatarsus adductus
  diagnosis, 258, 259–260, 265
  pathophysiology, 260
  stretching your infant's foot, 280
  treatment, 260
Metatarsus primus varus
  diagnosis, 258, 260
  treatment, 260
Methocarbamol (Robaxin), 27*t*
Minor tranquilizers, 26–27
Monoarthritis, differential diagnosis, 6
Morton's foot, vs Morton's neuroma, 244*n*
Morton's neuroma
  diagnosis, 244, 292*f*
  vs. Morton's foot, 244*n*
  pathophysiology, 244
  site of tenderness, 243*f*
  treatment, 244–245, 297*t*
"Movie sign", 183
Muscle
  cramps
    diagnosis, 311–312
    treatment, 312
  definition, 3
  diffuse pain
    evaluation, 310
    fibrositis, 310–311
    polymyalgia rheumatica, 311
  pull
    diagnosis, 307–308
    pathophysiology, 308
    treatment, 308
  tightness, and overuse injuries, 287
  trauma, 307–309
Muscle relaxants, 26, 27*t*
Muscle spasm theory, 126–127
Mycotic arthritis, 331
Myelography, 12

Myeloma, 139
Myofascial pain syndrome. *See* Fibrositis
Myositis ossificans
  definition, 4
  diagnosis, 309
  pathophysiology, 309
  treatment, 309
Myositis ossificans circumscripta, 309*n*
Myositis ossificans progressive, 309*n*
Myositis ossificans traumatica, 309*n*

Naproxen, 23*t*
Navicular (scaphoid) fracture, 88*f*, 94
NCV (nerve conduction velocity), 13
Neck. *See also* Upper back
  evaluation
    examination, 33
    history, 31–33
  exercises for patient, 43
  pain syndrome, 34–36*f*
    diagnosis, 32
    therapeutic measures, 37*t*
  posture of patient, 41
  relaxation, 42
  self care for patient, 39
  trauma
    acute strain, 37
    cervical sprain, 37–38
  unilateral pain, 296
  using a cervical collar, 40
Neoplasm(s)
  benign, and low back pain, 139
  of joint, 6
  primary malignant, and low back pain, 139
Nerve compression syndrome, definition, 4
Nerve conduction velocity (NCV), 13
Nerve root
  impingement. *See* Radiculopathy
  inflammation. *See* Radiculitis
Nerve syndromes, common peripheral
  lower extremity, 152*f*–153*f*
  upper extremity, 47*f*
Nerve(s), pinched. *See* Cervical radiculitis
Neural foramen, 115*f*
Neurologic symptoms and diffuse pain. *See also* specific syndromes
  of lower extremity
    evaluation, 150–155*f*
    syndromes, 155–158
  of upper extremity
    evaluation, 44–51*f*
    syndromes, 51–56
Neuroma, 108
Non-articular rheumatism. *See* Fibrositis
Norflex (orphenadrine), 27*t*
Norgesic, 27*t*
Norgesic forte, 27*t*
Nucleus pulposus, 115*f*
Nursemaid's elbow. *See* Radial head subluxation

Olecranon bursa, 78*f*
Olecranon bursitis
  diagnosis, 81
  pathophysiology, 81
  treatment, 81
Oligoarthritis, differential diagnosis, 6
Orphenadrine (Norflex), 27*t*
Ortolani's test, 260, 261*f*, 262
Osgood-Schlatter's disease, 3
  diagnosis, 268, 269
  pathophysiology, 269
  site of tenderness, 174*f*, 175*f*
  treatment, 269
Osteitis condensans, 138*n*
Osteitis pubis, 138*n*
Osteoarthritis. *See also* Degenerative arthritis
  clinical features, 315
  of elbow, 79–80
  of glenohumeral joint, 69–70
  of hand and fingers, 98
  patient handout, 318–319
Osteoblasts, 3
Osteochondritis dissecans, 3
  definition, 4
  diagnosis, 182, 185, 192, 193, 269
  pathophysiology, 193
  treatment, 193
Osteoclasts, 3
Osteocytes, 2
Osteomyelitis
  diagnosis, 138, 272, 333
  pathophysiology, 139
  treatment, 139
Osteoporosis
  patient handout, 337
  senile/postmenopausal
    diagnosis, 334–335
    pathophysiology, 335
    treatment, 335–336

INDEX

Overuse injuries, runner's. *See* Running injuries
Oxyphenbutazone, 24*t*

Pain. *See also* specific area
  chronic, 7–8
  determination of origin, 5–6
  differential diagnosis, 5–6
Pain medications, 26
Paraflex (chlorzoxazone), 27*t*
Parafon forte, 27*t*
Paralumbar muscles, 126
Paraspinal muscles, 126
Paronychia, 108*f*
Pars interarticularis stress fracture
  diagnosis, 117, 136, 272
  pathophysiology, 136
  treatment, 136
Patella(r)
  apprehension, 177*f*
  compression, 177*f*
  contusion
    diagnosis, 197–198
    pathophysiology, 198
    treatment, 198
  dislocation, acute
    diagnosis, 195, 197
    pathophysiology, 197
    treatment, 197
  inhibition, 177*f*
  subluxation, recurrent
    diagnosis, 176, 182, 185, 186–187, 191, 268
    pathophysiology, 187
    treatment, 187
  tendonitis
    diagnosis, 189, 295
    pathophysiology, 189–190
    site of tenderness, 174*f*
    treatment, 190
  tracking, 178
Patellofemoral pain
  in children, 268, 295
  diagnosis, 176, 182, 183, 185, 191, 192
  pathophysiology, 185
  patient handout, 199–200
  treatment, 185–186
Patellofemoral test, 177*f*
Pathologic fracture, definition, 4–5
Patient handout
  controlling pronation, 306
  decreasing shock, 305
  degenerative arthritis (osteoarthritis), 318–319
  healing your running injury, 304
  how to tape your ankle, 228
  low back exercises, 145
  neck exercises, 43
  osteoporosis, 337
  patellofemoral pain, 199–200
  preventing running injuries, 300
  rehabilitation exercises for epicondylitis, 83
  relaxation, 42
  shoulder mobilization exercises, 72–73
  ski safety, 204–205
  stretching, 301–303
  taking care of calluses and corns, 255
  tennis elbow, 84–85
  what to look for in a shoe, 254
  using a cervical collar, 40
  using heat and cold, 29–30
  if you keep spraining your ankle, 226–227
  your low back: posture and precautions, 143–144
  your low back-self care, 141–142
  your neck: posture, 41
  your neck: self care, 39
  your sprained ankle, 224–225
Patrick's test, 121*f*
Pauciarticular arthritis, 278
Pectoralis minor syndrome
  diagnosis, 51, 52
  pathophysiology, 52
  treatment, 52–53
Pediatric problems, 256–258
  abnormal curvature of back
    evaluation, 273
    structural scoliosis, 273–275*f*
  back pain in children and adolescents
    evaluation, 272
    Scheuermann's disease, 272–273
  bowleg, knocknee and backknee
    evaluation and management, 267–268*f*
  congenital foot problems
    examination, 256–257
    metatarses adductus, 259–260
    metatarses primus varus, 260
  congenital hip dysplasia
    congenital subluxability of hip, 262
    evaluation, 260–262*f*
  elbow problems in children, 275

Little League elbow, 276–277
radial head subluxation, 276
heel cord pain in children, 265
hip pain in children and adolescents
  evaluation, 269–270
  Legg-Perthe's disease, 271
  slipped capital femoral epiphysis, 271–272
  transient synovitis of hip, 270–271
intoeing gait
  evaluation, 265–266f
  excessive femoral anteversion, 267
  internal tibial torsion, 267
knee pain in children and adolescents
  evaluation, 268–269
  Osgood-Schlatter's disease, 269
musculoskeletal trauma in children, 277
  torus fractures of the distal radius, 278
patient handouts
  stretching your infant's foot, 280
  your baby's or child's shoes, 281
pronated foot in infants and children
  evaluation, 262–263
  flexible pronated foot, 263
systemic illnesses
  juvenile rheumatoid arthritis, 278
  rheumatic fever, 279
wryneck in infants and children, 275
Pedicle, 115f
Pendulum exercise, 72
Periarticular pain, vs. articular or referred pain, 5–6
Periosteum, 3
Periostitis, definition, 5
Peroneal nerve syndrome
  common, 157
  deep, 157
Peroneal tendonitis
  diagnosis, 213, 293
  pathophysiology, 213–214
  site of tenderness, 211f
  treatment, 297t
Pes cavus, 234f
Pes planus, 234f. *See* Foot strain in the hyperpronating foot, chronic
Phalanx, distal tuft fracture
  diagnosis, 100
  management, 101f
Phalen's maneuver, 50f, 51, 53
Phenylbutazone, 23t, 24t
Phosphorus, serum, 15
Physical therapy, 19–20

Pinched nerve. *See* Cervical radiculitis
PIP joint
  sprain
    central extensor slip tear, 104–105
    collateral ligament sprain, 105
    evaluation, 103
  dislocation
    diagnosis, 101
    extension splint for, 106f
    management, 101–102
Piroxicam, 23t
Pivot shift test, 176, 180f
Plantar fascia, 231f
Plantar fasciitis
  diagnosis, 236–237, 292f
  injection for, 239f
  pathophysiology, 238
  site of tenderness, 238f
  treatment, 238, 297t
Plantar wart
  diagnosis, 249f, 250
  pathophysiology, 250
  treatment, 250
Plica syndrome
  diagnosis, 187
  pathophysiology, 187
  treatment, 187
Podagra, 325
Polyarthritis, differential diagnosis, 320
Polyarticular arthritis, 278
Polymyalgia rheumatica
  clinical features and diagnosis, 311
  diagnosis, 45
  pathophysiology and treatment, 311
Polyneuropathy, 45, 150–151
Posterior calcaneal bursitis
  diagnosis, 221
  pathophysiology, 221
  treatment, 221
Posterior cruciate ligament injury
  diagnosis, 197
  treatment, 197
Posterior disc bulge, 137f
Posterior tibialis tendonitis
  diagnosis, 212, 293
  pathophysiology, 212
  site of tenderness, 211f
  treatment, 213, 297t
Postural abnormalities, 119f
Postural low back pain
  diagnosis, 117, 129

INDEX

Postural low back pain *(continued)*
  pathophysiology, 129
  treatment, 129, 143-144
Postural stress, 127
Posture, of neck, 41
Precepts of musculoskeletal management, 1
Prepatellar bursa swelling, 172*f*, 193
Prepatellar bursitis
  diagnosis, 193-194
  pathophysiology, 194
  treatment, 194
Prevention of further injury, 17
Probenecid, 327*t*
Progressive resistance exercises, short arc quadriceps, 202
Pronation of foot, 232
  controlling, 236, 306
  in infant and children, 259*f*, 262-263
Pronator teres syndrome
  diagnosis, 55
  pathophysiology, 55
  treatment, 55
Protein electrophoresis, serum, 15
Protrusion, posterior, 128
Pseudoclaudication
  diagnosis, 118, 151, 155-156
  pathophysiology, 156
  treatment, 156
Pseudogout
  clinical features and diagnosis, 326, 328
  pathophysiology, 328
  treatment, 328
Psoriatic arthritis
  clinical features, 322-323
  of hand and fingers, diagnosis, 98
  pathophysiology, 323
  treatment, 323
Psoriatic spondylitis, HLA-B27 antigen and, 15
Psychoactive medications, 26-28
Psychogenic muscle tension, in neck pain syndrome, 35
Pulled chest muscle
  diagnosis, 109-110
  treatment, 110
Pulled elbow. *See* Radial head subluxation
Pump bump, 221
Pyriformis syndrome
  diagnosis, 131, 135
  pathophysiology, 135
  treatment, 136

Q-angle, 182, 185*f*
Quadriceps
  angle, 182, 185*f*
  exercises, 201, 202
  stretching, 302

Radial head subluxation
  diagnosis, 276
  pathophysiology, 276
  treatment, 276
Radial nerve palsy
  diagnosis, 56
  treatment, 56
Radial nerve syndrome, diagnosis, 46*f*
Radial tunnel syndrome, diagnosis, 79
Radicular pain, 7
Radicular signs, 122*f*-123*f*
Radiculitis, definition, 5. *See also* specific type of radiculitis
Radiculopathy, definition, 5
Radius, distal torus fractures
  diagnosis, 278
  treatment, 278
Range of motion, 10
  elbow, 76
  hip, 162*f*
  knee, 175
  low back, 118
  neck exercises, 43
  shoulder, 62*f*
Raynaud's phenomenon, diagnosis, 48, 98
Reduction, of PIP joint dislocation, 102*f*
Referred pain
  in back of children and adolescents, 272
  in elbow, 74
  in hip, 159
  in knee, 168
  in low back, 114
  in shoulder, 57
  vs articular or periarticular pain, 5-6
"Reflex myalgia", 34-35, 64
Reflexes, upper extremity, 48*f*
Reiter's syndrome
  clinical features, 323-324
  HLA-B27 antigen and, 15
  pathophysiology, 324
  treatment, 324
Rela (carisoprodol), 27*t*
Relaxation techniques, 28, 42
Rest, 16-17
Retrocalcaneal bursitis
  diagnosis, 221

pathophysiology, 221
treatment, 221
Rheumatic diseases. *See also* specific diseases
  monoarthritis and, 6
Rheumatic fever
  clinical features, 279
  Jones' criteria for diagnosis, 279*t*
  pathophysiology, 279
  treatment, 279
Rheumatoid arthritis
  antinuclear antibodies and, 15
  clinical features, 321
  of hand and fingers, diagnosis, 98–99
  monoarthritis and, 6
  oligoarthritis and, 6
  pathophysiology, 321
  treatment, 322
Rheumatoid diseases
  diagnosis, 320
  enteropathic arthritis, 323
  other arthritides, 324
  psoriatic arthritis, 322–323
  Reiter's syndrome, 323–324
  rheumatoid arthritis, 321–322
  systemic lupus erythematosis, 322
Rheumatoid factor, 14–15
Rib bruise or fracture
  diagnosis, 111
  treatment, 111–112
Robaxin (methocarbamol), 27*t*
Robaxisal, 27*t*
Rotator cuff tears
  diagnosis, 60, 65, 68–69
  treatment, 66, 69
Rotator cuff tendonitis
  diagnosis, 60, 64–65
  injection, 67*f*
  sites of tenderness, 61*f*
  treatment, 66
Runners knee, 295
Runner's toe
  diagnosis, 291
  pathophysiology, 291
  treatment, 291
Running, biomechanics of, 289
Running injuries, 282–283
  controlling pronation, 306
  decreasing shock, 305
  differential diagnosis, 291–296*f*
  evaluation

examination, 289
history, 288
grades of overuse injury, 284
healing your, 304
hyperpronation-related overuse syndromes, 289*t*
overuse injuries, 283
preventing, 285, 300
principles of treatment of overuse injuries
  allow healing, 286
  correct stress concentrators, 287–288
  decrease inflammation, 286–287
runners toe, 291
skin problems, 289–291*f*
stress fractures, 284–285
stretching, 301–303
toe problems, 291
training, 284
treatment, 296*t*–299*t*

S1 root syndrome, 153*f*
Sacroileitis
  diagnosis, 137–138*f*
  treatment, 138
Sacroiliac joint signs, 125*f*
Salicylates, nonasprin, 22*t*
Salter's classification of epiphyseal injury, 258*t*
Sarcoidosis, 324
Scalenus anticus syndrome
  diagnosis, 50*f*, 52
  pathophysiology, 52
  treatment, 52–53
Scanogram, 120
Scaphoid (navicular) fracture, 88*f*, 94
Scheuermann's disease
  diagnosis, 272–273
  pathophysiology, 273
  treatment, 273
Sciatic notch, site of tenderness, 161*f*
Sciatica
  diagnosis, 131, 295
  pathophysiology, 131
  treatment, 132
Scleroderma, 15
Scoliosis
  diagnosis, 118, 119*f*, 128
  with lumbar derangement, 131

INDEX

Scoliosis *(continued)*
  structural
    diagnosis, 273, 274f
    pathophysiology, 275
    treatment, 275
Sedimentation rate, 14
Segmental instability syndrome,
    diagnosis, 117, 134
Septic arthritis
  diagnosis, 329–330
  pathophysiology, 330
  treatment, 330
Septic joint, 6
Serum protein electrophoresis, 15
Serum sickness, 324
Sesamoid bone
  fractures of
    diagnosis, 252
    treatment, 252
  hand lumps and, 107
Sesamoiditis, 240f
  diagnosis, 242, 291, 292f
  pathophysiology, 242
  treatment, 242, 297t
Sever's disease, 3
  diagnosis, 265
  pathophysiology, 265
  treatment, 265
Shin pain
  diagnosis, 293, 294f
  treatment of runners injuries, 298t
Shinsplints, diagnosis, 293
Shock, decreasing during running, 305
Shoes
  wear patterns in runners, 290f
  what to look for in, 254
  your baby's or child's, 281
Shoulder
  anatomy, 58f, 59f
  aspiration or injection technique, 69f
  degenerative arthritis, 60
  instability, 70
  mobilization exercises, 72–73
  pain and limitation of motion, 57–60
    examination, 60–63f
    history, 60
    process of, 65
  pointer, 71
  separation, 71
  shrugs, 43
  signs, 63f

  strain, 71
  tendonitis
    diagnosis, 64–65
    pathophysiology, 65
    treatment, 66
  trauma, 70–71
Sitting
  precautions for low back pain, 143
  prolonged, 127
Sjogren's syndrome, 15, 321
Ski safety, 204–205
Skin problems, for runnners, 289–291f
Sleeping, precautions for low back pain, 143
Slipped capital femoral epiphysis
  diagnosis, 271–272
  pathophysiology, 272
  treatment, 272
Snapping finger, 99
"Snuff box", tenderness of, 88f
Soft tissue infection. *See* Infections, soft tissue
Soma (carisoprodol), 27t
Spinal claudication. *See* Pseudoclaudication
Spinal cord lesion, diagnosis, 45, 150
Spinal stenosis. *See* Pseudoclaudication
Spinous process, 115f
Splint
  mallet finger, 107f
  for PIP joint, 105f
  for tuft fracture, 101f
Spondylitis, diagnosis, 118, 119
Spondylolisthesis, 118, 128f
  diagnosis, 137f, 272
  pathophysiology, 137
  treatment, 137
Spondylolysis, 128f. *See* Stress fracture, of pars interarticularis
Sprain, definition, 5. *See also* specific type of sprain
Spur(s)
  heel, 236–237
  lumbosacral, 115f
  toenail, 247f
Spurling's maneuver, 50f, 51
Spurs, 115f
Standing, precautions for low back pain, 144
Steinman's test, 174f
Sternoclavicular joint

anatomy, 111f
pain
  diagnosis, 110
  pathophysiology, 110
  treatment, 110
Steroid infection, 317. *See also* Injection technique
  for anserine tendonitis, 190f
  carpal tunnel, 54f
  for cuff tendonitis and subacromial bursitis, 67f
  of elbow, 80f
  for epicondylitis, 78f
  into first MTP joint, 242f
  for Morton's neuroma, 245f
  for plantar fasciitis, 239f
  for trochanteric bursitis, 164f
  patient instructions after injection, 25–26
  possible deleterious effects, 24–25
  precautions before, 24
  risks, 24
  technique, 25
Steroids
  disadvantages, 21–22
  injectable, 25t
Still's disease, 278
Stone bruise, 238–239
Straight leg raises, 122f, 123f, 202
Strain. *See* specific strain
Strength testing
  lower extremity, 155t
  for shoulder, 64t
Strengthening exercises
  abdominal, 145
  objectives, 18
Stress fracture
  definition, 5
  diagnosis, 293
  of femur, diagnosis, 295
  from running, 284–285
  of pars interarticularis
    diagnosis, 117, 136, 272
    pathophysiology, 136
    treatment, 136
  of metatarsal bone, diagnosis, 245
Stress injury
  definition, 5
  diagnosis, 293
Stretching, 20, 301–303
Subacromial bursa, 65

Subacromial bursitis
  diagnosis, 64–65
  injection, 67f
  sites of tenderness, 61f
  treatment, 66
Subdeltoid bursitis
  diagnosis, 64–65
  pathophysiology, 65
  treatment, 66
Subluxation
  definition, 5
  recurrent shoulder, 70
Subtalar joint
  degenerative arthritis
    diagnosis, 214–215
    pathophysiology, 215
    treatment, 215
  synovitis, treatment, 297t
  tenderness, 211f
Sulindac, 23t
Sulsinpyrazone, 327t
Supination, of foot, 233
Swelling, 10
Synovial fluid analysis, 3, 13–14t
Synovial membrane, 3
Synovial plica, inflammed, 185
Syphilitic arthritis, 331
Systemic lupus erythematosis, 322

Taping
  of ankle,
  of foot, 251f
  toes, 253f
Tardy ulnar palsy
  diagnosis, 55
  pathophysiology, 55
  treatment, 55
Tarsal tunnel syndrome
  diagnosis, 157–158, 232
  pathophysiology, 158
  treatment, 158
Taylor's bunion, 240
Temporomandibular joint pain, 338–340
Tenderness, localization of, 10
Tendon, definition, 3
Tendonitis. *See also* specific tendonitis
  definition, 5
  diagnosis, 182
Tennis elbow (lateral epicondylitis), 76
  patient handout, 84–85

rehabilitation exercises, 83
treatment, 78–79
Tenosynovitis, diagnosis and treatment, 332
TENS. *See* Transcutaneous nerve stimulator
Tensor facia lata strain, 295
Tetanus prophylaxis, 8*t*
Therapeutics. *See also* specific condition, treatment
  antiinflammatory medications, 20–22*t*
  exercise, 17–18
  heat and cold, 18–19, 18–19
  muscle relaxants, 26
  pain medications, 26
  physical therapy, 19–20, 19–20
  psychoactive medications and relaxation training, 26–28
  rest and immobilization, 16–17
  steroid injection, 000, 22, 24–26
Thigh
  circumference, 171
  pain in, 295
    evaluation, 159–160
    history, 160–161
    treatment of runner's injuries, 299*t*
Thomas' test, 270*f*
Thoracic outlet maneuvers, 50*f*, 51
Thoracic outlet syndromes
  diagnosis, 45, 52
  pathophysiology, 52
  treatment, 52–53
Throwing, overuse injuries related to, 276–277
Thumb
  degenerative arthritis of carpometacarpal joint, 90–91
  MCP joint sprains, 102–103
  trigger
    diagnosis, 99
    pathophysiology, 99
    treatment, 99
Tibial stress fracture, diagnosis, 293
Tibial torsion
  diagnosis, 266*f*
  treatment, 267
Tibialis tendonitis, anterior
  diagnosis, 214, 293
  pathophysiology, 214

site of tenderness, 211*f*
treatment, 214, 297*f*
Tinel's sign, 53
TMJ syndrome
  diagnosis, 338–339
  pathophysiology, 339
  treatment, 339–340
Toenails
  ingrown, 247–248
  proper trimming, 249*f*
Toe(s)
  dislocation of IP joint, treatment, 253*t*
  fracture
    diagnosis, 252
    treatment, 252
  hammertoes, 246
  ingrown toenail, 247–248
  nail spur, 247*f*
  problems in runners, 291
  trauma, 250–253*f*
Tolmetin, 23*t*
Tophi, 325
Torticollis
  acute (wryneck), 31
  diagnosis, 33–34
  pathophysiology, 33
  pediatric, 275
  treatment, 34
  congenital muscular, diagnosis, 275
Torus fractures of distal radius, diagnosis, 277, 278
Tranquilizers, minor, 26–27
Transcutaneous nerve stimulator (TENS), 20
Transient synovitis of hip
  diagnosis, 270
  pathophysiology, 270
  treatment, 271
Trauma
  chest, 110–112
  elbow, 81–82
  foot and toe, 250–253*f*
  knee, 195–198
  lower back, 127–128
  musculoskeletal in children, 277–278
  neck and upper back, 37–38
  shoulder, 70–71
  wrist, 94
Trigger finger
  diagnosis, 99

injection, 100f
pathophysiology, 99
treatment, 99
Trigger points, 310
injection, 36f
Trochanteric bursitis
diagnosis, 163, 295
injection, 164f
pathophysiology, 163–164
site of tenderness, 161f
treatment, 164
Tuberculous arthritis, 331
Tuft fracture of distal phalanx
diagnosis, 100
management, 101f
Twisted knee. *See* Collateral ligament sprain

Ulnar collateral ligament of MCP joint of thumb, sprain, 103, 104f
Ulnar nerve compression
diagnosis, 46f
at elbow. *See* Tardy ulnar palsy
at wrist
diagnosis, 56
pathophysiology, 56
treatment, 56
Ulnar neuritis. *See* Tardy ulnar palsy
Ultrasound, 20
Upper back. *See also* Neck
acute torticollis, 33–34
evaluation
examination, 33
history, 31–33
trauma, 37–38
Upper extremity. *See also* specific syndromes
examination, 44–51f
reflexes, 48f
syndromes, 51–56
Uric acid, 15
Urinalysis, 125

Vascular disease, diagnosis, 150
Vascular status, 10
Vastus medialis bulge, 176f
Vertebral body, 115f
Viral arthritis, 6
diagnosis, 331
treatment, 331

Volar plate injury, 102
diagnosis, 103
management, 103–104

Walking, biomechanics, 232–233
Wall climbing, 73
Warmth, 10
Weaver's bottom (ischeal bursitis), 164
Whiplash
diagnosis, 32, 37–38
treatment, 38
Williams' exercises, 127
Wilson's fracture, 103
Wright's maneuver, 50f
Wrist
anatomy, 87f
aspiration or injection technique, 92f
degenerative arthritis
diagnosis, 91
pathophysiology, 91
treatment, 92
lumps, 92–93
median nerve compression in. *See* Carpal tunnel syndrome
pain
examination, 87–88
history, 86–87
sites of tenderness, 88f
splint, 89, 90f
sprain
diagnosis, 94
treatment, 94
tendonitis
diagnosis, 89
pathophysiology, 89
treatment, 89
trauma, 94
evaluation, 94
ulnar nerve compression, 56
Wryneck. *See* Torticollis, acute

X-rays, 11
of ankle sprain, 217–218
cervical spine and chest, 51
elbow, fat pad sign on, 82f
hip, 163
low back, 120, 122, 124
of lower extremity, 155
of stress fracture, 285